INTERNATIONAL
CODE COUNCIL®

D0569402

2012

INTERNATIONAL
ENERGY CONSERVATION
CODE®

A Member of the International Code Family®

2IECC®

Become a **Building Safety Professional Member**
and Learn More about the Code Council

GO TO WWW.ICCSAFE.ORG for All Your Technical and
Professional Needs Including:

❯ Codes, Standards and Guidelines

❯ Membership Benefits

❯ Education and Certification

❯ Communications on Industry News

2012 International Energy Conservation Code®

First Printing: May 2011
Second Printing: August 2011
Third Printing: August 2013
Fourth Printing: October 2013
Fifth Printing: December 2015

ISBN: 978-1-60983-058-8 (soft-cover edition)

COPYRIGHT © 2011
by
INTERNATIONAL CODE COUNCIL, INC.

ALL RIGHTS RESERVED. This 2012 *International Energy Conservation Code*® is a copyrighted work owned by the International Code Council, Inc. Without advance written permission from the copyright owner, no part of this book may be reproduced, distributed or transmitted in any form or by any means, including, without limitation, electronic, optical or mechanical means (by way of example, and not limitation, photocopying, or recording by or in an information storage retrieval system). For information on permission to copy material exceeding fair use, please contact: Publications, 4051 West Flossmoor Road, Country Club Hills, IL 60478. Phone 1-888-ICC-SAFE (422-7233).

Trademarks: "International Code Council," the "International Code Council" logo and the "International Energy Conservation Code" are trademarks of the International Code Council, Inc.

PREFACE

Introduction

Internationally, code officials recognize the need for a modern, up-to-date energy conservation code addressing the design of energy-efficient building envelopes and installation of energy efficient mechanical, lighting and power systems through requirements emphasizing performance. The *International Energy Conservation Code®*, in this 2012 edition, is designed to meet these needs through model code regulations that will result in the optimal utilization of fossil fuel and nondepletable resources in all communities, large and small.

This code contains separate provisions for commercial buildings and for low-rise residential buildings (3 stories or less in height above grade.) Each set of provisions in this code—IECC-Commercial Provisions and IECC—Residential Provisions are separately applied to buildings within their respective scopes. Each set of provisions are to be treated separately; they each contain a Scope and Administration chapter, a Definitions chapter, a General Requirements chapter, and a chapter containing energy efficiency requirements applicable to buildings within their scope.

This comprehensive energy conservation code establishes minimum regulations for energy efficient buildings using prescriptive and performance-related provisions. It is founded on broad-based principles that make possible the use of new materials and new energy efficient designs. This 2012 edition is fully compatible with all of the *International Codes®* (I-Codes®) published by the International Code Council (ICC)®, including: the *International Building Code®, International Existing Building Code®, International Fire Code®, International Fuel Gas Code®, International Green Construction Code™* (to be available March 2012), *International Mechanical Code®*, ICC *Performance Code®, International Plumbing Code®, International Private Sewage Disposal Code®, International Property Maintenance Code®, International Residential Code®, International Swimming Pool and Spa Code™* (to be available March 2012), *International Wildland-Urban Interface Code®* and *International Zoning Code®*.

The *International Energy Conservation Code* provisions provide many benefits, among which is the model code development process that offers an international forum for energy professionals to discuss performance and prescriptive code requirements. This forum provides an excellent arena to debate proposed revisions. This model code also encourages international consistency in the application of provisions.

Development

The first edition of the *International Energy Conservation Code* (1998) was based on the 1995 edition of the *Model Energy Code* promulgated by the Council of American Building Officials (CABO) and included changes approved through the CABO Code Development Procedures through 1997. CABO assigned all rights and responsibilities to the International Code Council and its three statutory members at that time, including Building Officials and Code Administrators International, Inc. (BOCA), International Conference of Building Officials (ICBO) and Southern Building Code Congress International (SBCCI). This 2012 edition presents the code as originally issued, with changes reflected in the 2000, 2003, 2006 and 2009 editions and further changes approved through the ICC Code Development Process through 2010. A new edition such as this is promulgated every three years.

This code is founded on principles intended to establish provisions consistent with the scope of an energy conservation code that adequately conserves energy; provisions that do not unnecessarily increase construction costs; provisions that do not restrict the use of new materials, products or methods of construction; and provisions that do not give preferential treatment to particular types or classes of materials, products or methods of construction.

Adoption

The *International Energy Conservation Code* is available for adoption and use by jurisdictions internationally. Its use within a governmental jurisdiction is intended to be accomplished through adoption by reference in accordance with proceedings establishing the jurisdiction's laws. At the time of adoption, jurisdictions should insert the appropriate information in provisions requiring specific local information, such as the name of the adopting jurisdiction. These locations are shown in bracketed words in small capital letters in the code and in the sample ordinance. The sample adoption ordinance on page ix addresses several key elements of a code adoption ordinance, including the information required for insertion into the code text.

Maintenance

The *International Energy Conservation Code* is kept up to date through the review of proposed changes submitted by code enforcement officials, industry representatives, design professionals and other interested parties. Proposed changes are carefully considered through an open code development process in which all interested and affected parties may participate.

The contents of this work are subject to change both through the Code Development Cycles and the governmental body that enacts the code into law. For more information regarding the code development process, contact the Codes and Standards Development Department of the International Code Council.

While the development procedure of the *International Energy Conservation Code* assures the highest degree of care, ICC, its members and those participating in the development of this code do not accept any liability resulting from compliance or noncompliance with the provisions because ICC and its members do not have the power or authority to police or enforce compliance with the contents of this code. Only the governmental body that enacts the code into law has such authority.

Code Development Committee Responsibilities
(Letter Designations in Front of Section Numbers)

In each code development cycle, proposed changes to the code are considered at the Code Development Hearings by the applicable International Code Development Committee. The IECC—Commercial Provisions (sections designated with a "C" prior to the section number) are primarily maintained by the Commercial Energy Code Development Committee. The IECC—Residential Provisions (sections designated with an "R" prior to the section number) are maintained by the Residential Energy Code Development Committee. This is designated in the chapter headings by a [CE] and [RE], respectively. Proposed changes to a code section or defined term, other than those designated by [CE] or [RE], that has a number beginning with a letter in brackets are considered by a different code development committee. For example, proposed changes to code sections or defined terms that have [M] in front of them are considered by the International Mechanical Code Development Committee.

Maintenance responsibilities for the IECC are designated as follows:

[CE] = Commercial Energy Code Development Committee

[M] = International Mechanical Code Development Committee

[RE] = Residential Energy Code Development Committee

Note that, for the development of the 2015 edition of the I-Codes, there will be two groups of code development committees and they will meet in separate years. The groupings are as follows:

Group A Codes (Heard in 2012, Code Change Proposals Deadline: January 3, 2012)	Group B Codes (Heard in 2013, Code Change Proposals Deadline: January 3, 2013)
International Building Code	Administrative Provisions (Chapter 1 all codes except the IECC, IRC and ICCPC, administrative updates to currently referenced standards, and designated definitions)
International Fuel Gas Code	International Energy Conservation Code
International Mechanical Code	International Existing Building Code
International Plumbing Code	International Fire Code
International Private Sewage Disposal Code	International Green Construction Code
	ICC Performance Code
	International Property Maintenance Code
	International Residential Code
	International Swimming Pool and Spa Code
	International Wildland-Urban Interface Code
	International Zoning Code

Code change proposals submitted for code sections that have a letter designation in front of them will be heard by the respective committee responsible for such code sections. Because different committees will meet in different years, it is possible that some proposals for this code will be heard by a committee in a different year than the year in which the primary committees for this code meets.

For example, the definition of the term "Energy Recovery Ventilation System" for the IECC—Commercial Provisions (page C-8) is the responsibility of the International Mechanical Code Development Committee, which is part of the Group A code hearings. Therefore, any proposed changes to this defined term will need to be submitted by the deadline for the Group A codes, so that the International Mechanical Code Development Committee can consider that proposed change during the 2012 Code Change Cycle.

It is very important that anyone submitting code change proposals understand which code development committee is responsible for the section of the code that is the subject of the code change proposal. For further information on the code development committee responsibilities, please visit the ICC web site at www.iccsafe.org/scoping.

Marginal Markings

Solid vertical lines in the margins within the body of the code indicate a technical change from the requirements of the 2009 edition. Deletion indicators in the form of an arrow (➡) are provided in the margin where an entire section, paragraph, exception or table has been deleted or an item in a list of items or a table has been deleted.

A single asterisk [*] placed in the margin indicates that text or a table has been relocated within the code. A double asterisk [**] placed in the margin indicates that the text or table immediately following it has been relocated there from elsewhere in the code. The following table indicates such relocations in the 2012 Edition of the *International Energy Conservation Code*.

2012 LOCATION	2009 LOCATION
C402.4.3	502.4.1
C408.2.2.1	503.2.9.1
C408.2.2.2	503.2.9.2
C408.2.5.2	503.2.9.3

Italicized Terms

Selected terms set forth in Chapter 2, Definitions, for both the Commercial and Residential Provisions are italicized where they appear in code text. Such terms are not italicized where the definition set forth in Chapter 2 does not impart the intended meaning in the use of the term. The terms selected have definitions which the user should read carefully to facilitate better understanding of the code.

Effective Use of the International Energy Conservation Code

The *International Energy Conservation Code* (IECC) is a model code that regulates minimum energy conservation requirements for new buildings. The IECC addresses energy conservation requirements for all aspects of energy uses in both commercial and residential construction, including heating and ventilating, lighting, water heating, and power usage for appliances and building systems.

The IECC is a design document. For example, before one constructs a building, the designer must determine the minimum insulation *R*-values and fenestration *U*-factors for the building exterior envelope. Depending on whether the building is for residential use or for commercial use, the IECC sets forth minimum requirements for exterior envelope insulation, window and door *U*-factors and SHGC ratings, duct insulation, lighting and power efficiency, and water distribution insulation.

Arrangement and Format of the 2012 IECC

The IECC contains two separate sets of provisions—one for commercial buildings and one for residential buildings. Each set of provisions are applied separately to buildings within their scope. The IECC—Commercial Provisions apply to all buildings except for residential buildings 3 stories or less in height. The IECC—Residential Provisions apply to detached one- and two-family dwellings and multiple single family dwellings as well as Group R-2, R-3 and R-4 buildings three stories or less in height. These scopes are based on the definitions of "commercial building" and "residential building," respectively, in Chapter 2 of each set of provisions. Note that the IECC—Commercial Provisions therefore contain provisions for residential buildings 4 stories or greater in height. Each set of provisions is divided into 4 different parts:

Chapters	Subjects
1-2	Administration and definitions
3	Climate zones and general materials requirements
4	Energy efficiency requirements
5	Referenced standards

The following is a chapter-by-chapter synopsis of the scope and intent of the provisions of the *International Energy Conservation Code* and applies to both the commercial and residential energy provisions:

Chapter 1 Administration. This chapter contains provisions for the application, enforcement and administration of subsequent requirements of the code. In addition to establishing the scope of the code, Chapter 1 identifies which buildings and structures come under its purview. Chapter 1 is largely concerned with maintaining "due process of law" in enforcing the energy conservation criteria contained in the body of the code. Only through careful observation of the administrative provisions can the building official reasonably expect to demonstrate that "equal protection under the law" has been provided.

Chapter 2 Definitions. All terms that are defined in the code are listed alphabetically in Chapter 2. While a defined term may be used in one chapter or another, the meaning provided in Chapter 2 is applicable throughout the code.

Additional definitions regarding climate zones are found in Tables 301.3(1) and (2). These are not listed in Chapter 2.

Where understanding of a term's definition is especially key to or necessary for understanding of a particular code provision, the term is shown in *italics* wherever it appears in the code. This is true only for those terms that have a meaning that is unique to the code. In other words, the generally understood meaning of a term or phrase might not be sufficient or consistent with the meaning prescribed by the code; therefore, it is essential that the code-defined meaning be known.

Guidance regarding tense, gender and plurality of defined terms as well as guidance regarding terms not defined in this code is provided.

Chapter 3 General Requirements. Chapter 3 specifies the climate zones that will serve to establish the exterior design conditions. In addition, Chapter 3 provides interior design conditions that are used as a basis for assumptions in heating and cooling load calculations, and provides basic material requirements for insulation materials and fenestration materials.

Climate has a major impact on the energy use of most buildings. The code establishes many requirements such as wall and roof insulation *R*-values, window and door thermal transmittance requirement (*U*-factors) as well as provisions that affect the mechanical systems based upon the climate where the building is located. This chapter contains information that will be used to properly assign the building location into the correct climate zone and is used as the basis for establishing requirements or elimination of requirements.

Chapter 4 Energy Efficiency. Chapter 4 of each set of provisions contains the technical requirements for energy efficiency.

Commercial Energy Efficiency. Chapter 4 of the IECC—Commercial Provisions contains the energy-efficiency-related requirements for the design and construction of most types of commercial buildings and residential buildings greater than three stories in height above grade. Residential buildings, townhouses and garden apartments three stories or less in height are covered in the IECC—Residential Provisions. This chapter defines requirements for the portions of the building and building systems that impact energy use in new commercial construction and new residential construction greater than three stories in height, and promotes the effective use of energy. The provisions within the chapter promote energy efficiency in the building envelope, the heating and cooling system and the service water heating system of the building.

Residential Energy Efficiency. Chapter 4 of the IECC—Residential Provisions contains the energy-efficiency-related requirements for the design and construction of residential buildings regulated under this code. It should be noted that the definition of a *residential building* in this code is unique for this code. In this code, a *residential building* is a detached one- and two-family dwelling and multiple single family dwellings as well as R-2, R-3 or R-4 building three stories or less in height. All other buildings, including residential buildings greater than three stories in height, are regulated by the energy conservation requirements in the IECC—Commercial Provisions. The applicable portions of a residential building must comply with the provisions within this chapter for energy efficiency. This chapter defines requirements for the portions of the building and building systems that impact energy use in new residential construction and promotes the effective use of energy. The provisions within the chapter promote energy efficiency in the building envelope, the heating and cooling system and the service water heating system of the building.

Chapter 5 Referenced Standards. The code contains numerous references to standards that are used to regulate materials and methods of construction. Chapter 5 contains a comprehensive list of all standards that are referenced in the code. The standards are part of the code to the extent of the reference to the standard. Compliance with the referenced standard is necessary for compliance with this code. By providing specifically adopted standards, the construction and installation requirements necessary for compliance with the code can be readily determined. The basis for code compliance is, therefore, established and available on an equal basis to the code official, contractor, designer and owner.

Chapter 5 is organized in a manner that makes it easy to locate specific standards. It lists all of the referenced standards, alphabetically, by acronym of the promulgating agency of the standard. Each agency's standards are then listed in either alphabetical or numeric order based upon the standard identification. The list also contains the title of the standard; the edition (date) of the standard referenced; any addenda included as part of the ICC adoption; and the section or sections of this code that reference the standard.

LEGISLATION

The International Codes are designed and promulgated to be adopted by reference by legislative action. Jurisdictions wishing to adopt the 2012 *International Energy Conservation Code* as an enforceable regulation governing energy efficient building envelopes and installation of energy efficient mechanical, lighting and power systems should ensure that certain factual information is included in the adopting legislation at the time adoption is being considered by the appropriate governmental body. The following sample adoption legislation addresses several key elements, including the information required for insertion into the code text.

SAMPLE LEGISLATION FOR ADOPTION OF THE *INTERNATIONAL ENERGY CONSERVATION CODE* ORDINANCE NO._____

A[N] [ORDINANCE/STATUTE/REGULATION] of the [JURISDICTION] adopting the 2012 edition of the *International Energy Conservation Code*, regulating and governing energy efficient building envelopes and installation of energy efficient mechanical, lighting and power systems in the [JURISDICTION]; providing for the issuance of permits and collection of fees therefor; repealing [ORDINANCE/STATUTE/REGULATION] No. _____ of the [JURISDICTION] and all other ordinances or parts of laws in conflict therewith.

The [GOVERNING BODY] of the [JURISDICTION] does ordain as follows:

Section 1. That a certain document, three (3) copies of which are on file in the office of the [TITLE OF JURISDICTION'S KEEPER OF RECORDS] of [NAME OF JURISDICTION], being marked and designated as the *International Energy Conservation Code*, 2012 edition, as published by the International Code Council, be and is hereby adopted as the Energy Conservation Code of the [JURISDICTION], in the State of [STATE NAME] for regulating and governing energy efficient building envelopes and installation of energy efficient mechanical, lighting and power systems as herein provided; providing for the issuance of permits and collection of fees therefor; and each and all of the regulations, provisions, penalties, conditions and terms of said Energy Conservation Code on file in the office of the [JURISDICTION] are hereby referred to, adopted, and made a part hereof, as if fully set out in this legislation, with the additions, insertions, deletions and changes, if any, prescribed in Section 2 of this ordinance.

Section 2. The following sections are hereby revised:

Sections C101.1 and R101.1. Insert: [NAME OF JURISDICTION].

Sections C108.4 and R108.4. Insert: [DOLLAR AMOUNT] in two places.

Section 3. That [ORDINANCE/STATUTE/REGULATION] No. _____ of [JURISDICTION] entitled [FILL IN HERE THE COMPLETE TITLE OF THE LEGISLATION OR LAWS IN EFFECT AT THE PRESENT TIME SO THAT THEY WILL BE REPEALED BY DEFINITE MENTION] and all other ordinances or parts of laws in conflict herewith are hereby repealed.

Section 4. That if any section, subsection, sentence, clause or phrase of this legislation is, for any reason, held to be unconstitutional, such decision shall not affect the validity of the remaining portions of this ordinance. The [GOVERNING BODY] hereby declares that it would have passed this law, and each section, subsection, clause or phrase thereof, irrespective of the fact that any one or more sections, subsections, sentences, clauses and phrases be declared unconstitutional.

Section 5. That nothing in this legislation or in the Energy Conservation Code hereby adopted shall be construed to affect any suit or proceeding impending in any court, or any rights acquired, or liability incurred, or any cause or causes of action acquired or existing, under any act or ordinance hereby repealed as cited in Section 3 of this law; nor shall any just or legal right or remedy of any character be lost, impaired or affected by this legislation.

Section 6. That the [JURISDICTION'S KEEPER OF RECORDS] is hereby ordered and directed to cause this legislation to be published. (An additional provision may be required to direct the number of times the legislation is to be published and to specify that it is to be in a newspaper in general circulation. Posting may also be required.)

Section 7. That this law and the rules, regulations, provisions, requirements, orders and matters established and adopted hereby shall take effect and be in full force and effect [TIME PERIOD] from and after the date of its final passage and adoption.

TABLE OF CONTENTS

IECC—COMMERCIAL PROVISIONS

TABLE OF CONTENTS

CHAPTER 1 [CE]

SCOPE AND ADMINISTRATION

PART 1—SCOPE AND APPLICATION

SECTION C101
SCOPE AND GENERAL REQUIREMENTS

C101.1 Title. This code shall be known as the *International Energy Conservation Code* of **[NAME OF JURISDICTION]**, and shall be cited as such. It is referred to herein as "this code."

C101.2 Scope. This code applies to *commercial buildings* and the buildings sites and associated systems and equipment.

C101.3 Intent. This code shall regulate the design and construction of buildings for the effective use and conservation of energy over the useful life of each building. This code is intended to provide flexibility to permit the use of innovative approaches and techniques to achieve this objective. This code is not intended to abridge safety, health or environmental requirements contained in other applicable codes or ordinances.

C101.4 Applicability. Where, in any specific case, different sections of this code specify different materials, methods of construction or other requirements, the most restrictive shall govern. Where there is a conflict between a general requirement and a specific requirement, the specific requirement shall govern.

C101.4.1 Existing buildings. Except as specified in this chapter, this code shall not be used to require the removal, *alteration* or abandonment of, nor prevent the continued use and maintenance of, an existing building or building system lawfully in existence at the time of adoption of this code.

C101.4.2 Historic buildings. Any building or structure that is listed in the State or National Register of Historic Places; designated as a historic property under local or state designation law or survey; certified as a contributing resource with a National Register listed or locally designated historic district; or with an opinion or certification that the property is eligible to be listed on the National or State Registers of Historic Places either individually or as a contributing building to a historic district by the State Historic Preservation Officer or the Keeper of the National Register of Historic Places, are exempt from this code.

C101.4.3 Additions, alterations, renovations or repairs. Additions, alterations, renovations or repairs to an existing building, building system or portion thereof shall conform to the provisions of this code as they relate to new construction without requiring the unaltered portion(s) of the existing building or building system to comply with this code. Additions, alterations, renovations or repairs shall not create an unsafe or hazardous condition or overload existing building systems. An addition shall be deemed to comply with this code if the addition alone complies or if the existing building and addition comply with this code as a single building.

Exception: The following need not comply provided the energy use of the building is not increased:

1. Storm windows installed over existing fenestration.
2. Glass only replacements in an existing sash and frame.
3. Existing ceiling, wall or floor cavities exposed during construction provided that these cavities are filled with insulation.
4. Construction where the existing roof, wall or floor cavity is not exposed.
5. Reroofing for roofs where neither the sheathing nor the insulation is exposed. Roofs without insulation in the cavity and where the sheathing or insulation is exposed during reroofing shall be insulated either above or below the sheathing.
6. Replacement of existing doors that separate *conditioned space* from the exterior shall not require the installation of a vestibule or revolving door, provided, however, that an existing vestibule that separates a *conditioned space* from the exterior shall not be removed,
7. Alterations that replace less than 50 percent of the luminaires in a space, provided that such alterations do not increase the installed interior lighting power.
8. Alterations that replace only the bulb and ballast within the existing luminaires in a space provided that the *alteration* does not increase the installed interior lighting power.

C101.4.4 Change in occupancy or use. Spaces undergoing a change in occupancy that would result in an increase in demand for either fossil fuel or electrical energy shall comply with this code. Where the use in a space changes from one use in Table C405.5.2(1) or (2) to another use in Table C405.5.2(1) or (2), the installed lighting wattage shall comply with Section C405.5.

C101.4.5 Change in space conditioning. Any nonconditioned space that is altered to become *conditioned space* shall be required to be brought into full compliance with this code.

C101.4.6 Mixed occupancy. Where a building includes both *residential* and *commercial* occupancies, each occupancy shall be separately considered and meet the applicable provisions of IECC—Commercial Provisions or IECC—Residential Provisions.

C101.5 Compliance. *Residential buildings* shall meet the provisions of IECC—Residential Provisions. *Commercial* *buildings* shall meet the provisions of IECC—Commercial Provisions.

C101.5.1 Compliance materials. The *code official* shall be permitted to approve specific computer software, worksheets, compliance manuals and other similar materials that meet the intent of this code.

C101.5.2 Low energy buildings. The following buildings, or portions thereof, separated from the remainder of the building by *building thermal envelope* assemblies complying with this code shall be exempt from the *building thermal envelope* provisions of this code:

1. Those with a peak design rate of energy usage less than 3.4 Btu/h · ft^2 (10.7 W/m^2) or 1.0 watt/ft^2 (10.7 W/m^2) of floor area for space conditioning purposes.

2. Those that do not contain *conditioned space.*

SECTION C102
ALTERNATE MATERIALS—METHOD OF CONSTRUCTION, DESIGN OR INSULATING SYSTEMS

C102.1 General. This code is not intended to prevent the use of any material, method of construction, design or insulating system not specifically prescribed herein, provided that such construction, design or insulating system has been *approved* by the *code official* as meeting the intent of this code.

C102.1.1 Above code programs. The *code official* or other authority having jurisdiction shall be permitted to deem a national, state or local energy efficiency program to exceed the energy efficiency required by this code. Buildings *approved* in writing by such an energy efficiency program shall be considered in compliance with this code. The requirements identified as "mandatory" in Chapter 4 shall be met.

PART 2—ADMINISTRATION AND ENFORCEMENT

SECTION C103
CONSTRUCTION DOCUMENTS

C103.1 General. Construction documents and other supporting data shall be submitted in one or more sets with each application for a permit. The construction documents shall be prepared by a registered design professional where required by the statutes of the jurisdiction in which the project is to be constructed. Where special conditions exist, the *code official* is authorized to require necessary construction documents to be prepared by a registered design professional.

Exception: The *code official* is authorized to waive the requirements for construction documents or other supporting data if the *code official* determines they are not necessary to confirm compliance with this code.

C103.2 Information on construction documents. Construction documents shall be drawn to scale upon suitable material. Electronic media documents are permitted to be submitted when *approved* by the *code official*. Construction documents shall be of sufficient clarity to indicate the location, nature and extent of the work proposed, and show in sufficient detail pertinent data and features of the building, systems and equipment as herein governed. Details shall include, but are not limited to, as applicable, insulation materials and their *R*-values; fenestration *U*-factors and SHGCs; area-weighted *U*-factor and SHGC calculations; mechanical system design criteria; mechanical and service water heating system and equipment types, sizes and efficiencies; economizer description; equipment and systems controls; fan motor horsepower (hp) and controls; duct sealing, duct and pipe insulation and location; lighting fixture schedule with wattage and control narrative; and air sealing details.

C103.3 Examination of documents. The *code official* shall examine or cause to be examined the accompanying construction documents and shall ascertain whether the construction indicated and described is in accordance with the requirements of this code and other pertinent laws or ordinances.

C103.3.1 Approval of construction documents. When the *code official* issues a permit where construction documents are required, the construction documents shall be endorsed in writing and stamped "Reviewed for Code Compliance." Such *approved* construction documents shall not be changed, modified or altered without authorization from the *code official*. Work shall be done in accordance with the *approved* construction documents.

One set of construction documents so reviewed shall be retained by the *code official*. The other set shall be returned to the applicant, kept at the site of work and shall be open to inspection by the *code official* or a duly authorized representative.

C103.3.2 Previous approvals. This code shall not require changes in the construction documents, construction or designated occupancy of a structure for which a lawful permit has been heretofore issued or otherwise lawfully authorized, and the construction of which has been pursued in good faith within 180 days after the effective date of this code and has not been abandoned.

C103.3.3 Phased approval. The *code official* shall have the authority to issue a permit for the construction of part of an energy conservation system before the construction documents for the entire system have been submitted or *approved*, provided adequate information and detailed statements have been filed complying with all pertinent requirements of this code. The holders of such permit shall proceed at their own risk without assurance that the permit for the entire energy conservation system will be granted.

C103.4 Amended construction documents. Changes made during construction that are not in compliance with the *approved* construction documents shall be resubmitted for approval as an amended set of construction documents.

C103.5 Retention of construction documents. One set of *approved* construction documents shall be retained by the *code official* for a period of not less than 180 days from date of completion of the permitted work, or as required by state or local laws.

SECTION C104
INSPECTIONS

C104.1 General. Construction or work for which a permit is required shall be subject to inspection by the *code official*.

C104.2 Required approvals. Work shall not be done beyond the point indicated in each successive inspection without first obtaining the approval of the *code official*. The *code official*, upon notification, shall make the requested inspections and shall either indicate the portion of the construction that is satisfactory as completed, or notify the permit holder or his or her agent wherein the same fails to comply with this code. Any portions that do not comply shall be corrected and such portion shall not be covered or concealed until authorized by the *code official*.

C104.3 Final inspection. The building shall have a final inspection and not be occupied until *approved*.

C104.4 Reinspection. A building shall be reinspected when determined necessary by the *code official*.

C104.5 Approved inspection agencies. The *code official* is authorized to accept reports of *approved* inspection agencies, provided such agencies satisfy the requirements as to qualifications and reliability.

C104.6 Inspection requests. It shall be the duty of the holder of the permit or their duly authorized agent to notify the *code official* when work is ready for inspection. It shall be the duty of the permit holder to provide access to and means for inspections of such work that are required by this code.

C104.7 Reinspection and testing. Where any work or installation does not pass an initial test or inspection, the necessary corrections shall be made so as to achieve compliance with this code. The work or installation shall then be resubmitted to the *code official* for inspection and testing.

C104.8 Approval. After the prescribed tests and inspections indicate that the work complies in all respects with this code, a notice of approval shall be issued by the *code official*.

C104.8.1 Revocation. The *code official* is authorized to, in writing, suspend or revoke a notice of approval issued under the provisions of this code wherever the certificate is issued in error, or on the basis of incorrect information supplied, or where it is determined that the building or structure, premise, or portion thereof is in violation of any ordinance or regulation or any of the provisions of this code.

SECTION C105
VALIDITY

C105.1 General. If a portion of this code is held to be illegal or void, such a decision shall not affect the validity of the remainder of this code.

SECTION C106
REFERENCED STANDARDS

C106.1 Referenced codes and standards. The codes and standards referenced in this code shall be those listed in Chapter 5, and such codes and standards shall be considered as part of the requirements of this code to the prescribed extent of each such reference and as further regulated in Sections C106.1.1 and C106.1.2.

C106.1.1 Conflicts. Where conflicts occur between provisions of this code and referenced codes and standards, the provisions of this code shall apply.

C106.1.2 Provisions in referenced codes and standards. Where the extent of the reference to a referenced code or standard includes subject matter that is within the scope of this code, the provisions of this code, as applicable, shall take precedence over the provisions in the referenced code or standard.

C106.2 Conflicting requirements. Where the provisions of this code and the referenced standards conflict, the provisions of this code shall take precedence.

C106.3 Application of references. References to chapter or section numbers, or to provisions not specifically identified by number, shall be construed to refer to such chapter, section or provision of this code.

C106.4 Other laws. The provisions of this code shall not be deemed to nullify any provisions of local, state or federal law.

SECTION C107
FEES

C107.1 Fees. A permit shall not be issued until the fees prescribed in Section C107.2 have been paid, nor shall an amendment to a permit be released until the additional fee, if any, has been paid.

C107.2 Schedule of permit fees. A fee for each permit shall be paid as required, in accordance with the schedule as established by the applicable governing authority.

C107.3 Work commencing before permit issuance. Any person who commences any work before obtaining the necessary permits shall be subject to an additional fee established by the *code official*, which shall be in addition to the required permit fees.

C107.4 Related fees. The payment of the fee for the construction, *alteration*, removal or demolition of work done in connection to or concurrently with the work or activity authorized by a permit shall not relieve the applicant or holder of the permit from the payment of other fees that are prescribed by law.

C107.5 Refunds. The *code official* is authorized to establish a refund policy.

SECTION C108
STOP WORK ORDER

C108.1 Authority. Whenever the *code official* finds any work regulated by this code being performed in a manner either contrary to the provisions of this code or dangerous or unsafe, the *code official* is authorized to issue a stop work order.

C108.2 Issuance. The stop work order shall be in writing and shall be given to the owner of the property involved, or to the owner's agent, or to the person doing the work. Upon issuance of a stop work order, the cited work shall immediately cease. The stop work order shall state the reason for the order, and the conditions under which the cited work will be permitted to resume.

C108.3 Emergencies. Where an emergency exists, the *code official* shall not be required to give a written notice prior to stopping the work.

C108.4 Failure to comply. Any person who shall continue any work after having been served with a stop work order, except such work as that person is directed to perform to remove a violation or unsafe condition, shall be liable to a fine of not less than [AMOUNT] dollars or more than [AMOUNT] dollars.

SECTION C109
BOARD OF APPEALS

C109.1 General. In order to hear and decide appeals of orders, decisions or determinations made by the *code official* relative to the application and interpretation of this code, there shall be and is hereby created a board of appeals. The *code official* shall be an ex officio member of said board but shall have no vote on any matter before the board. The board of appeals shall be appointed by the governing body and shall hold office at its pleasure. The board shall adopt rules of procedure for conducting its business, and shall render all decisions and findings in writing to the appellant with a duplicate copy to the *code official*.

C109.2 Limitations on authority. An application for appeal shall be based on a claim that the true intent of this code or the rules legally adopted thereunder have been incorrectly interpreted, the provisions of this code do not fully apply or an equally good or better form of construction is proposed. The board shall have no authority to waive requirements of this code.

C109.3 Qualifications. The board of appeals shall consist of members who are qualified by experience and training and are not employees of the jurisdiction.

CHAPTER 2 [CE]

DEFINITIONS

SECTION C201
GENERAL

C201.1 Scope. Unless stated otherwise, the following words and terms in this code shall have the meanings indicated in this chapter.

C201.2 Interchangeability. Words used in the present tense include the future; words in the masculine gender include the feminine and neuter; the singular number includes the plural and the plural includes the singular.

C201.3 Terms defined in other codes. Terms that are not defined in this code but are defined in the *International Building Code, International Fire Code, International Fuel Gas Code, International Mechanical Code, International Plumbing Code* or the *International Residential Code* shall have the meanings ascribed to them in those codes.

C201.4 Terms not defined. Terms not defined by this chapter shall have ordinarily accepted meanings such as the context implies.

SECTION C202
GENERAL DEFINITIONS

ACCESSIBLE. Admitting close approach as a result of not being guarded by locked doors, elevation or other effective means (see "Readily *accessible*").

ADDITION. An extension or increase in the *conditioned space* floor area or height of a building or structure.

AIR BARRIER. Material(s) assembled and joined together to provide a barrier to air leakage through the building envelope. An air barrier may be a single material or a combination of materials.

ALTERATION. Any construction or renovation to an existing structure other than repair or addition that requires a permit. Also, a change in a mechanical system that involves an extension, addition or change to the arrangement, type or purpose of the original installation that requires a permit.

APPROVED. Approval by the *code official* as a result of investigation and tests conducted by him or her, or by reason of accepted principles or tests by nationally recognized organizations.

AUTOMATIC. Self-acting, operating by its own mechanism when actuated by some impersonal influence, as, for example, a change in current strength, pressure, temperature or mechanical configuration (see "Manual").

BUILDING. Any structure used or intended for supporting or sheltering any use or occupancy, including any mechanical systems, service water heating systems and electric power and lighting systems located on the building site and supporting the building.

BUILDING COMMISSIONING. A process that verifies and documents that the selected building systems have been designed, installed, and function according to the owner's project requirements and construction documents, and to minimum code requirements.

BUILDING ENTRANCE. Any door, set of doors, doorway, or other form of portal that is used to gain access to the building from the outside by the public.

BUILDING SITE. A contiguous area of land that is under the ownership or control of one entity.

BUILDING THERMAL ENVELOPE. The basement walls, exterior walls, floor, roof, and any other building elements that enclose *conditioned space* or provides a boundary between *conditioned space* and exempt or unconditioned space.

C-FACTOR (THERMAL CONDUCTANCE). The coefficient of heat transmission (surface to surface) through a building component or assembly, equal to the time rate of heat flow per unit area and the unit temperature difference between the warm side and cold side surfaces (Btu/h ft^2 × °F) [W/(m^2 × K)].

CODE OFFICIAL. The officer or other designated authority charged with the administration and enforcement of this code, or a duly authorized representative.

COEFFICENT OF PERFORMANCE (COP) – COOLING. The ratio of the rate of heat removal to the rate of energy input, in consistent units, for a complete refrigerating system or some specific portion of that system under designated operating conditions.

COEFFICIENT OF PERFORMANCE (COP) – HEATING. The ratio of the rate of heat delivered to the rate of energy input, in consistent units, for a complete heat pump system, including the compressor and, if applicable, auxiliary heat, under designated operating conditions.

COMMERCIAL BUILDING. For this code, all buildings that are not included in the definition of "Residential buildings."

CONDITIONED FLOOR AREA. The horizontal projection of the floors associated with the *conditioned space*.

CONDITIONED SPACE. An area or room within a building being heated or cooled, containing uninsulated ducts, or with a fixed opening directly into an adjacent *conditioned space*.

CONTINUOUS AIR BARRIER. A combination of materials and assemblies that restrict or prevent the passage of air through the building thermal envelope.

CRAWL SPACE WALL. The opaque portion of a wall that encloses a crawl space and is partially or totally below grade.

CURTAIN WALL. Fenestration products used to create an external nonload-bearing wall that is designed to separate the exterior and interior environments.

DAYLIGHT ZONE.

1. **Under skylights.** The area under skylights whose horizontal dimension, in each direction, is equal to the skylight dimension in that direction plus either the floor-to-ceiling height or the dimension to a ceiling height opaque partition, or one-half the distance to adjacent skylights or vertical fenestration, whichever is least.

2. **Adjacent to vertical fenestration.** The area adjacent to vertical fenestration which receives daylight through the fenestration. For purposes of this definition and unless more detailed analysis is provided, the daylight *zone* depth is assumed to extend into the space a distance of 15 feet (4572 mm) or to the nearest ceiling height opaque partition, whichever is less. The daylight *zone* width is assumed to be the width of the window plus 2 feet (610 mm) on each side, or the window width plus the distance to an opaque partition, or the window width plus one-half the distance to adjacent skylight or vertical fenestration, whichever is least.

DEMAND CONTROL VENTILATION (DCV). A ventilation system capability that provides for the automatic reduction of outdoor air intake below design rates when the actual occupancy of spaces served by the system is less than design occupancy.

DEMAND RECIRCULATION WATER SYSTEM. A water distribution system where pump(s) prime the service hot water piping with heated water upon demand for hot water.

DUCT. A tube or conduit utilized for conveying air. The air passages of self-contained systems are not to be construed as air ducts.

DUCT SYSTEM. A continuous passageway for the transmission of air that, in addition to ducts, includes duct fittings, dampers, plenums, fans and accessory air-handling equipment and appliances.

[B] DWELLING UNIT. A single unit providing complete independent living facilities for one or more persons, including permanent provisions for living, sleeping, eating, cooking and sanitation.

DYNAMIC GLAZING. Any fenestration product that has the fully reversible ability to change its performance properties, including *U*-factor, SHGC, or VT.

ECONOMIZER, AIR. A duct and damper arrangement and automatic control system that allows a cooling system to supply outside air to reduce or eliminate the need for mechanical cooling during mild or cold weather.

ECONOMIZER, WATER. A system where the supply air of a cooling system is cooled indirectly with water that is itself cooled by heat or mass transfer to the environment without the use of mechanical cooling.

ENCLOSED SPACE. A volume surrounded by solid surfaces such as walls, floors, roofs, and openable devices such as doors and operable windows.

ENERGY ANALYSIS. A method for estimating the annual energy use of the *proposed design* and *standard reference design* based on estimates of energy use.

ENERGY COST. The total estimated annual cost for purchased energy for the building functions regulated by this code, including applicable demand charges.

[M] ENERGY RECOVERY VENTILATION SYSTEM. Systems that employ air-to-air heat exchangers to recover energy from exhaust air for the purpose of preheating, precooling, humidifying or dehumidifying outdoor ventilation air prior to supplying the air to a space, either directly or as part of an HVAC system.

ENERGY SIMULATION TOOL. An *approved* software program or calculation-based methodology that projects the annual energy use of a building.

ENTRANCE DOOR. Fenestration products used for ingress, egress and access in nonresidential buildings, including, but not limited to, exterior entrances that utilize latching hardware and automatic closers and contain over 50-percent glass specifically designed to withstand heavy use and possibly abuse.

EQUIPMENT ROOM. A space that contains either electrical equipment, mechanical equipment, machinery, water pumps or hydraulic pumps that are a function of the building's services.

EXTERIOR WALL. Walls including both above-grade walls and basement walls.

FAN BRAKE HORSEPOWER (BHP). The horsepower delivered to the fan's shaft. Brake horsepower does not include the mechanical drive losses (belts, gears, etc.).

FAN SYSTEM BHP. The sum of the fan brake horsepower of all fans that are required to operate at fan system design conditions to supply air from the heating or cooling source to the *conditioned space(s)* and return it to the source or exhaust it to the outdoors.

FAN SYSTEM DESIGN CONDITIONS. Operating conditions that can be expected to occur during normal system operation that result in the highest supply fan airflow rate to conditioned spaces served by the system.

FAN SYSTEM MOTOR NAMEPLATE HP. The sum of the motor nameplate horsepower of all fans that are required to operate at design conditions to supply air from the heating or cooling source to the *conditioned space(s)* and return it to the source or exhaust it to the outdoors.

FENESTRATION. Skylights, roof windows, vertical windows (fixed or moveable), opaque doors, glazed doors, glazed block and combination opaque/glazed doors. Fenestration includes products with glass and nonglass glazing materials.

FENESTRATION PRODUCT, FIELD-FABRICATED. A fenestration product whose frame is made at the construction site of standard dimensional lumber or other materials that were not previously cut, or otherwise formed with the specific intention of being used to fabricate a fenestration product or exterior door. Field fabricated does not include site-built fenestration.

FENESTRATION PRODUCT, SITE-BUILT. A fenestration designed to be made up of field-glazed or field-assembled units using specific factory cut or otherwise factory-formed framing and glazing units. Examples of site-built fenestration include storefront systems, curtain walls, and atrium roof systems.

_F_-FACTOR. The perimeter heat loss factor for slab-on-grade floors (Btu/h × ft × °F) [W/(m × K)].

FURNACE ELECTRICITY RATIO. The ratio of furnace electricity use to total furnace energy computed as ratio = $(3.412 \times E_{AE})/1000 \times E_F + 3.412 \times E_{AE})$ where E_{AE} (average annual auxiliary electrical consumption) and E_F (average annual fuel energy consumption) are defined in Appendix N to Subpart B of Part 430 of Title 10 of the Code of Federal Regulations and E_F is expressed in millions of Btu's per year.

GENERAL LIGHTING. Lighting that provides a substantially uniform level of illumination throughout an area. General lighting shall not include decorative lighting or lighting that provides a dissimilar level of illumination to serve a specialized application or feature within such area.

HEAT TRAP. An arrangement of piping and fittings, such as elbows, or a commercially available heat trap that prevents thermosyphoning of hot water during standby periods.

HEATED SLAB. Slab-on-grade construction in which the heating elements, hydronic tubing, or hot air distribution system is in contact with, or placed within or under, the slab.

HIGH-EFFICACY LAMPS. Compact fluorescent lamps, T-8 or smaller diameter linear fluorescent lamps, or lamps with a minimum efficacy of:

1. 60 lumens per watt for lamps over 40 watts;
2. 50 lumens per watt for lamps over 15 watts to 40 watts; and
3. 40 lumens per watt for lamps 15 watts or less.

HUMIDISTAT. A regulatory device, actuated by changes in humidity, used for automatic control of relative humidity.

INFILTRATION. The uncontrolled inward air leakage into a building caused by the pressure effects of wind or the effect of differences in the indoor and outdoor air density or both.

INSULATING SHEATHING. An insulating board with a core material having a minimum _R_-value of R-2.

INTEGRATED PART LOAD VALUE (IPLV). A single-number figure of merit based on part-load EER, COP, or kW/ton expressing part-load efficiency for air-conditioning and heat pump equipment on the basis of weighted operation at various load capacities for equipment.

LABELED. Equipment, materials or products to which have been affixed a label, seal, symbol or other identifying mark of a nationally recognized testing laboratory, inspection agency or other organization concerned with product evaluation that maintains periodic inspection of the production of the above-labeled items and whose labeling indicates either that

the equipment, material or product meets identified standards or has been tested and found suitable for a specified purpose.

LISTED. Equipment, materials, products or services included in a list published by an organization acceptable to the _code official_ and concerned with evaluation of products or services that maintains periodic inspection of production of _listed_ equipment or materials or periodic evaluation of services and whose listing states either that the equipment, material, product or service meets identified standards or has been tested and found suitable for a specified purpose.

LOW-VOLTAGE LIGHTING. Lighting equipment powered through a transformer such as a cable conductor, a rail conductor and track lighting.

MANUAL. Capable of being operated by personal intervention (see "Automatic").

NAMEPLATE HORSEPOWER. The nominal motor horsepower rating stamped on the motor nameplate.

NONSTANDARD PART LOAD VALUE (NPLV). A single-number part-load efficiency figure of merit calculated and referenced to conditions other than IPLV conditions, for units that are not designed to operate at ARI standard rating conditions.

ON-SITE RENEWABLE ENERGY. Energy derived from solar radiation, wind, waves, tides, landfill gas, biomass, or the internal heat of the earth. The energy system providing on-site renewable energy shall be located on the project site.

PROPOSED DESIGN. A description of the proposed building used to estimate annual energy use for determining compliance based on total building performance.

READILY ACCESSIBLE. Capable of being reached quickly for operation, renewal or inspection without requiring those to whom ready access is requisite to climb over or remove obstacles or to resort to portable ladders or access equipment (see "_Accessible_").

REPAIR. The reconstruction or renewal of any part of an existing building.

RESIDENTIAL BUILDING. For this code, includes detached one- and two-family dwellings and multiple single-family dwellings (townhouses) as well as Group R-2, R-3 and R-4 buildings three stories or less in height above grade plane.

ROOF ASSEMBLY. A system designed to provide weather protection and resistance to design loads. The system consists of a roof covering and roof deck or a single component serving as both the roof covering and the roof deck. A roof assembly includes the roof covering, underlayment, roof deck, insulation, vapor retarder and interior finish.

_R_-VALUE (THERMAL RESISTANCE). The inverse of the time rate of heat flow through a body from one of its bounding surfaces to the other surface for a unit temperature difference between the two surfaces, under steady state conditions, per unit area ($h \cdot ft^2 \cdot °F/Btu$) [($m^2 \cdot K$)/W].

SCREW LAMP HOLDERS. A lamp base that requires a screw-in-type lamp, such as a compact-fluorescent, incandescent, or tungsten-halogen bulb.

SERVICE WATER HEATING. Supply of hot water for purposes other than comfort heating.

SKYLIGHT. Glass or other transparent or translucent glazing material installed at a slope of less than 60 degrees (1.05 rad) from horizontal. Glazing material in skylights, including unit skylights, solariums, sunrooms, roofs and sloped walls is included in this definition.

[B] SLEEPING UNIT. A room or space in which people sleep, which can also include permanent provisions for living, eating, and either sanitation or kitchen facilities but not both. Such rooms and spaces that are also part of a dwelling unit are not *sleeping units*.

SOLAR HEAT GAIN COEFFICIENT (SHGC). The ratio of the solar heat gain entering the space through the fenestration assembly to the incident solar radiation. Solar heat gain includes directly transmitted solar heat and absorbed solar radiation which is then reradiated, conducted or convected into the space.

STANDARD REFERENCE DESIGN. A version of the *proposed design* that meets the minimum requirements of this code and is used to determine the maximum annual energy use requirement for compliance based on total building performance.

STOREFRONT. A nonresidential system of doors and windows mulled as a composite fenestration structure that has been designed to resist heavy use. *Storefront* systems include, but are not limited to, exterior fenestration systems that span from the floor level or above to the ceiling of the same story on commercial buildings, with or without mulled windows and doors.

SUNROOM. A one-story structure attached to a dwelling with a glazing area in excess of 40 percent of the gross area of the structure's exterior walls and roof.

THERMAL ISOLATION. Physical and space conditioning separation from *conditioned space(s)*. The *conditioned space*(s) shall be controlled as separate zones for heating and cooling or conditioned by separate equipment.

THERMOSTAT. An automatic control device used to maintain temperature at a fixed or adjustable set point.

***U*-FACTOR (THERMAL TRANSMITTANCE).** The coefficient of heat transmission (air to air) through a building component or assembly, equal to the time rate of heat flow per unit area and unit temperature difference between the warm side and cold side air films (Btu/h · ft^2 · °F) [W/(m^2 · K)].

[M] VENTILATION. The natural or mechanical process of supplying conditioned or unconditioned air to, or removing such air from, any space.

[M] VENTILATION AIR. That portion of supply air that comes from outside (outdoors) plus any recirculated air that has been treated to maintain the desired quality of air within a designated space.

VISIBLE TRANSMITTANCE [VT]. The ratio of visible light entering the space through the fenestration product assembly to the incident visible light, Visible Transmittance, includes the effects of glazing material and frame and is expressed as a number between 0 and 1.

ZONE. A space or group of spaces within a building with heating or cooling requirements that are sufficiently similar so that desired conditions can be maintained throughout using a single controlling device.

GENERAL REQUIREMENTS

SECTION C301
CLIMATE ZONES

C301.1 General. Climate zones from Figure C301.1 or Table C301.1 shall be used in determining the applicable requirements from Chapter 4. Locations not in Table C301.1 (outside the United States) shall be assigned a climate zone based on Section C301.3.

C301.2 Warm humid counties. Warm humid counties are identified in Table C301.1 by an asterisk.

C301.3 International climate zones. The climate zone for any location outside the United States shall be determined by applying Table C301.3(1) and then Table C301.3(2).

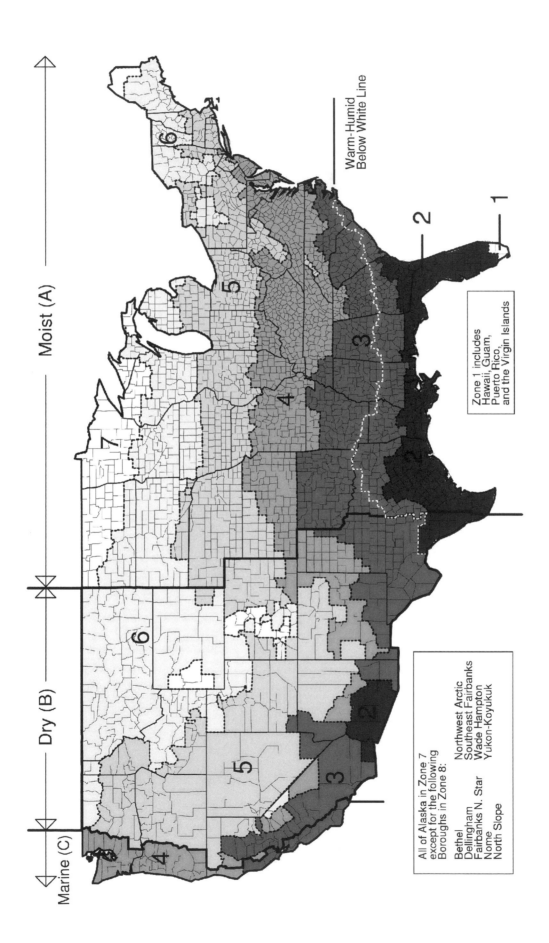

**FIGURE C301.1
CLIMATE ZONES**

Moist (A)

Dry (B)

Marine (C)

Warm-Humid
Below White Line

Zone 1 includes
Hawaii, Guam,
Puerto Rico,
and the Virgin Islands

All of Alaska in Zone 7
except for the following
Boroughs in Zone 8:

Bethel   Northwest Arctic
Dellingham   Southeast Fairbanks
Fairbanks N. Star Wade Hampton
Nome   Yukon-Koyukuk
North Slope

TABLE C301.1
CLIMATE ZONES, MOISTURE REGIMES, AND WARM-HUMID
DESIGNATIONS BY STATE, COUNTY AND TERRITORY

Key: A – Moist, B – Dry, C – Marine. Absence of moisture designation indicates moisture regime is irrelevant.
Asterisk (*) indicates a warm-humid location.

US STATES

ALABAMA

3A Autauga*
2A Baldwin*
3A Barbour*
3A Bibb
3A Blount
3A Bullock*
3A Butler*
3A Calhoun
3A Chambers
3A Cherokee
3A Chilton
3A Choctaw*
3A Clarke*
3A Clay
3A Cleburne
3A Coffee*
3A Colbert
3A Conecuh*
3A Coosa
3A Covington*
3A Crenshaw*
3A Cullman
3A Dale*
3A Dallas*
3A DeKalb
3A Elmore*
3A Escambia*
3A Etowah
3A Fayette
3A Franklin
3A Geneva*
3A Greene
3A Hale
3A Henry*
3A Houston*
3A Jackson
3A Jefferson
3A Lamar
3A Lauderdale
3A Lawrence

3A Lee
3A Limestone
3A Lowndes*
3A Macon*
3A Madison
3A Marengo*
3A Marion
3A Marshall
2A Mobile*
3A Monroe*
3A Montgomery*
3A Morgan
3A Perry*
3A Pickens
3A Pike*
3A Randolph
3A Russell*
3A Shelby
3A St. Clair
3A Sumter
3A Talladega
3A Tallapoosa
3A Tuscaloosa
3A Walker
3A Washington*
3A Wilcox*
3A Winston

ALASKA

7 Aleutians East
7 Aleutians West
7 Anchorage
8 Bethel
7 Bristol Bay
7 Denali
8 Dillingham
8 Fairbanks North Star
7 Haines
7 Juneau
7 Kenai Peninsula
7 Ketchikan Gateway

7 Kodiak Island
7 Lake and Peninsula
7 Matanuska-Susitna
8 Nome
8 North Slope
8 Northwest Arctic
7 Prince of Wales
  Outer Ketchikan
7 Sitka
7 Skagway-Hoonah-
  Angoon
8 Southeast Fairbanks
7 Valdez-Cordova
8 Wade Hampton
7 Wrangell-Petersburg
7 Yakutat
8 Yukon-Koyukuk

ARIZONA

5B Apache
3B Cochise
5B Coconino
4B Gila
3B Graham
3B Greenlee
2B La Paz
2B Maricopa
3B Mohave
5B Navajo
2B Pima
2B Pinal
3B Santa Cruz
4B Yavapai
2B Yuma

ARKANSAS

3A Arkansas
3A Ashley
4A Baxter
4A Benton
4A Boone
3A Bradley

3A Calhoun
4A Carroll
3A Chicot
3A Clark
3A Clay
3A Cleburne
3A Cleveland
3A Columbia*
3A Conway
3A Craighead
3A Crawford
3A Crittenden
3A Cross
3A Dallas
3A Desha
3A Drew
3A Faulkner
3A Franklin
4A Fulton
3A Garland
3A Grant
3A Greene
3A Hempstead*
3A Hot Spring
3A Howard
3A Independence
4A Izard
3A Jackson
3A Jefferson
3A Johnson
3A Lafayette*
3A Lawrence
3A Lee
3A Lincoln
3A Little River*
3A Logan
3A Lonoke
4A Madison
4A Marion
3A Miller*
3A Mississippi

3A Monroe
3A Montgomery
3A Nevada
4A Newton
3A Ouachita
3A Perry
3A Phillips
3A Pike
3A Poinsett
3A Polk
3A Pope
3A Prairie
3A Pulaski
3A Randolph
3A Saline
3A Scott
4A Searcy
3A Sebastian
3A Sevier*
3A Sharp
3A St. Francis
4A Stone
3A Union*
3A Van Buren
4A Washington
3A White
3A Woodruff
3A Yell

CALIFORNIA

3C Alameda
6B Alpine
4B Amador
3B Butte
4B Calaveras
3B Colusa
3B Contra Costa
4C Del Norte
4B El Dorado
3B Fresno
3B Glenn

(continued)

TABLE C301.1—continued
CLIMATE ZONES, MOISTURE REGIMES, AND WARM-HUMID
DESIGNATIONS BY STATE, COUNTY AND TERRITORY

4C Humboldt
2B Imperial
4B Inyo
3B Kern
3B Kings
4B Lake
5B Lassen
3B Los Angeles
3B Madera
3C Marin
4B Mariposa
3C Mendocino
3B Merced
5B Modoc
6B Mono
3C Monterey
3C Napa
5B Nevada
3B Orange
3B Placer
5B Plumas
3B Riverside
3B Sacramento
3C San Benito
3B San Bernardino
3B San Diego
3C San Francisco
3B San Joaquin
3C San Luis Obispo
3C San Mateo
3C Santa Barbara
3C Santa Clara
3C Santa Cruz
3B Shasta
5B Sierra
5B Siskiyou
3B Solano
3C Sonoma
3B Stanislaus
3B Sutter
3B Tehama
4B Trinity
3B Tulare
4B Tuolumne
3C Ventura
3B Yolo

3B Yuba

COLORADO

5B Adams
6B Alamosa
5B Arapahoe
6B Archuleta
4B Baca
5B Bent
5B Boulder
6B Chaffee
5B Cheyenne
7 Clear Creek
6B Conejos
6B Costilla
5B Crowley
6B Custer
5B Delta
5B Denver
6B Dolores
5B Douglas
6B Eagle
5B Elbert
5B El Paso
5B Fremont
5B Garfield
5B Gilpin
7 Grand
7 Gunnison
7 Hinsdale
5B Huerfano
7 Jackson
5B Jefferson
5B Kiowa
5B Kit Carson
7 Lake
5B La Plata
5B Larimer
4B Las Animas
5B Lincoln
5B Logan
5B Mesa
7 Mineral
6B Moffat
5B Montezuma
5B Montrose

5B Morgan
4B Otero
6B Ouray
7 Park
5B Phillips
7 Pitkin
5B Prowers
5B Pueblo
6B Rio Blanco
7 Rio Grande
7 Routt
6B Saguache
7 San Juan
6B San Miguel
5B Sedgwick
7 Summit
5B Teller
5B Washington
5B Weld
5B Yuma

CONNECTICUT

5A (all)

DELAWARE

4A (all)

DISTRICT OF COLUMBIA

4A (all)

FLORIDA

2A Alachua*
2A Baker*
2A Bay*
2A Bradford*
2A Brevard*
1A Broward*
2A Calhoun*
2A Charlotte*
2A Citrus*
2A Clay*
2A Collier*
2A Columbia*
2A DeSoto*
2A Dixie*
2A Duval*

2A Escambia*
2A Flagler*
2A Franklin*
2A Gadsden*
2A Gilchrist*
2A Glades*
2A Gulf*
2A Hamilton*
2A Hardee*
2A Hendry*
2A Hernando*
2A Highlands*
2A Hillsborough*
2A Holmes*
2A Indian River*
2A Jackson*
2A Jefferson*
2A Lafayette*
2A Lake*
2A Lee*
2A Leon*
2A Levy*
2A Liberty*
2A Madison*
2A Manatee*
2A Marion*
2A Martin*
1A Miami-Dade*
1A Monroe*
2A Nassau*
2A Okaloosa*
2A Okeechobee*
2A Orange*
2A Osceola*
2A Palm Beach*
2A Pasco*
2A Pinellas*
2A Polk*
2A Putnam*
2A Santa Rosa*
2A Sarasota*
2A Seminole*
2A St. Johns*
2A St. Lucie*
2A Sumter*
2A Suwannee*

2A Taylor*
2A Union*
2A Volusia*
2A Wakulla*
2A Walton*
2A Washington*

GEORGIA

2A Appling*
2A Atkinson*
2A Bacon*
2A Baker*
3A Baldwin
4A Banks
3A Barrow
3A Bartow
3A Ben Hill*
2A Berrien*
3A Bibb
3A Bleckley*
2A Brantley*
2A Brooks*
2A Bryan*
3A Bulloch*
3A Burke
3A Butts
3A Calhoun*
2A Camden*
3A Candler*
3A Carroll
4A Catoosa
2A Charlton*
2A Chatham*
3A Chattahoochee*
4A Chattooga
3A Cherokee
3A Clarke
3A Clay*
3A Clayton
2A Clinch*
3A Cobb
3A Coffee*
2A Colquitt*
3A Columbia
2A Cook*
3A Coweta

(continued)

2012 INTERNATIONAL ENERGY CONSERVATION CODE®

TABLE C301.1—continued
CLIMATE ZONES, MOISTURE REGIMES, AND WARM-HUMID
DESIGNATIONS BY STATE, COUNTY AND TERRITORY

3A Crawford	2A Lanier*	3A Taylor*	5B Cassia	4A Crawford
3A Crisp*	3A Laurens*	3A Telfair*	6B Clark	5A Cumberland
4A Dade	3A Lee*	3A Terrell*	5B Clearwater	5A DeKalb
4A Dawson	2A Liberty*	2A Thomas*	6B Custer	5A De Witt
2A Decatur*	3A Lincoln	3A Tift*	5B Elmore	5A Douglas
3A DeKalb	2A Long*	2A Toombs*	6B Franklin	5A DuPage
3A Dodge*	2A Lowndes*	4A Towns	6B Fremont	5A Edgar
3A Dooly*	4A Lumpkin	3A Treutlen*	5B Gem	4A Edwards
3A Dougherty*	3A Macon*	3A Troup	5B Gooding	4A Effingham
3A Douglas	3A Madison	3A Turner*	5B Idaho	4A Fayette
3A Early*	3A Marion*	3A Twiggs*	6B Jefferson	5A Ford
2A Echols*	3A McDuffie	4A Union	5B Jerome	4A Franklin
2A Effingham*	2A McIntosh*	3A Upson	5B Kootenai	5A Fulton
3A Elbert	3A Meriwether	4A Walker	5B Latah	4A Gallatin
3A Emanuel*	2A Miller*	3A Walton	6B Lemhi	5A Greene
2A Evans*	2A Mitchell*	2A Ware*	5B Lewis	5A Grundy
4A Fannin	3A Monroe	3A Warren	5B Lincoln	4A Hamilton
3A Fayette	3A Montgomery*	3A Washington	6B Madison	5A Hancock
4A Floyd	3A Morgan	2A Wayne*	5B Minidoka	4A Hardin
3A Forsyth	4A Murray	3A Webster*	5B Nez Perce	5A Henderson
4A Franklin	3A Muscogee	3A Wheeler*	6B Oneida	5A Henry
3A Fulton	3A Newton	4A White	5B Owyhee	5A Iroquois
4A Gilmer	3A Oconee	4A Whitfield	5B Payette	4A Jackson
3A Glascock	3A Oglethorpe	3A Wilcox*	5B Power	4A Jasper
2A Glynn*	3A Paulding	3A Wilkes	5B Shoshone	4A Jefferson
4A Gordon	3A Peach*	3A Wilkinson	6B Teton	5A Jersey
2A Grady*	4A Pickens	3A Worth*	5B Twin Falls	5A Jo Daviess
3A Greene	2A Pierce*	**HAWAII**	6B Valley	4A Johnson
3A Gwinnett	3A Pike		5B Washington	5A Kane
4A Habersham	3A Polk	1A (all)*	**ILLINOIS**	5A Kankakee
4A Hall	3A Pulaski*	**IDAHO**		5A Kendall
3A Hancock	3A Putnam		5A Adams	5A Knox
3A Haralson	3A Quitman*	5B Ada	4A Alexander	5A Lake
3A Harris	4A Rabun	6B Adams	4A Bond	5A La Salle
3A Hart	3A Randolph*	6B Bannock	5A Boone	4A Lawrence
3A Heard	3A Richmond	6B Bear Lake	5A Brown	5A Lee
3A Henry	3A Rockdale	5B Benewah	5A Bureau	5A Livingston
3A Houston*	3A Schley*	6B Bingham	5A Calhoun	5A Logan
3A Irwin*	3A Screven*	6B Blaine	5A Carroll	5A Macon
3A Jackson	2A Seminole*	6B Boise	5A Cass	4A Macoupin
3A Jasper	3A Spalding	6B Bonner	5A Champaign	4A Madison
2A Jeff Davis*	4A Stephens	6B Bonneville	4A Christian	4A Marion
3A Jefferson	3A Stewart*	6B Boundary	5A Clark	5A Marshall
3A Jenkins*	3A Sumter*	6B Butte	4A Clay	5A Mason
3A Johnson*	3A Talbot	6B Camas	4A Clinton	5A Massac
3A Jones	3A Taliaferro	5B Canyon	5A Coles	5A McDonough
3A Lamar	2A Tattnall*	6B Caribou	5A Cook	5A McHenry

(continued)

TABLE C301.1—continued
CLIMATE ZONES, MOISTURE REGIMES, AND WARM-HUMID
DESIGNATIONS BY STATE, COUNTY AND TERRITORY

5A McLean	5A Boone	5A Miami	5A Appanoose	5A Jasper
5A Menard	4A Brown	4A Monroe	5A Audubon	5A Jefferson
5A Mercer	5A Carroll	5A Montgomery	5A Benton	5A Johnson
4A Monroe	5A Cass	5A Morgan	6A Black Hawk	5A Jones
4A Montgomery	4A Clark	5A Newton	5A Boone	5A Keokuk
5A Morgan	5A Clay	5A Noble	6A Bremer	6A Kossuth
5A Moultrie	5A Clinton	4A Ohio	6A Buchanan	5A Lee
5A Ogle	4A Crawford	4A Orange	6A Buena Vista	5A Linn
5A Peoria	4A Daviess	5A Owen	6A Butler	5A Louisa
4A Perry	4A Dearborn	5A Parke	6A Calhoun	5A Lucas
5A Piatt	5A Decatur	4A Perry	5A Carroll	6A Lyon
5A Pike	5A De Kalb	4A Pike	5A Cass	5A Madison
4A Pope	5A Delaware	5A Porter	5A Cedar	5A Mahaska
4A Pulaski	4A Dubois	4A Posey	6A Cerro Gordo	5A Marion
5A Putnam	5A Elkhart	5A Pulaski	6A Cherokee	5A Marshall
4A Randolph	5A Fayette	5A Putnam	6A Chickasaw	5A Mills
4A Richland	4A Floyd	5A Randolph	5A Clarke	6A Mitchell
5A Rock Island	5A Fountain	4A Ripley	6A Clay	5A Monona
4A Saline	5A Franklin	5A Rush	6A Clayton	5A Monroe
5A Sangamon	5A Fulton	4A Scott	5A Clinton	5A Montgomery
5A Schuyler	4A Gibson	5A Shelby	5A Crawford	5A Muscatine
5A Scott	5A Grant	4A Spencer	5A Dallas	6A O'Brien
4A Shelby	4A Greene	5A Starke	5A Davis	6A Osceola
5A Stark	5A Hamilton	5A Steuben	5A Decatur	5A Page
4A St. Clair	5A Hancock	5A St. Joseph	6A Delaware	6A Palo Alto
5A Stephenson	4A Harrison	4A Sullivan	5A Des Moines	6A Plymouth
5A Tazewell	5A Hendricks	4A Switzerland	6A Dickinson	6A Pocahontas
4A Union	5A Henry	5A Tippecanoe	5A Dubuque	5A Polk
5A Vermilion	5A Howard	5A Tipton	6A Emmet	5A Pottawattamie
4A Wabash	5A Huntington	5A Union	6A Fayette	5A Poweshiek
5A Warren	4A Jackson	4A Vanderburgh	6A Floyd	5A Ringgold
4A Washington	5A Jasper	5A Vermillion	6A Franklin	6A Sac
4A Wayne	5A Jay	5A Vigo	5A Fremont	5A Scott
4A White	4A Jefferson	5A Wabash	5A Greene	5A Shelby
5A Whiteside	4A Jennings	5A Warren	6A Grundy	6A Sioux
5A Will	5A Johnson	4A Warrick	5A Guthrie	5A Story
4A Williamson	4A Knox	4A Washington	6A Hamilton	5A Tama
5A Winnebago	5A Kosciusko	5A Wayne	6A Hancock	5A Taylor
5A Woodford	5A Lagrange	5A Wells	6A Hardin	5A Union
	5A Lake	5A White	5A Harrison	5A Van Buren
INDIANA	5A La Porte	5A Whitley	5A Henry	5A Wapello
	4A Lawrence		6A Howard	5A Warren
5A Adams	5A Madison	**IOWA**	6A Humboldt	5A Washington
5A Allen	5A Marion		6A Ida	5A Wayne
5A Bartholomew	5A Marshall	5A Adair	5A Iowa	6A Webster
5A Benton	4A Martin	5A Adams	5A Jackson	6A Winnebago
5A Blackford		6A Allamakee		

(continued)

TABLE C301.1—continued
CLIMATE ZONES, MOISTURE REGIMES, AND WARM-HUMID
DESIGNATIONS BY STATE, COUNTY AND TERRITORY

6A Winneshiek	4A Haskell	4A Sedgwick	2A Iberville*	6A Cumberland
5A Woodbury	4A Hodgeman	4A Seward	3A Jackson*	6A Franklin
6A Worth	4A Jackson	4A Shawnee	2A Jefferson*	6A Hancock
6A Wright	4A Jefferson	5A Sheridan	2A Jefferson Davis*	6A Kennebec
KANSAS	5A Jewell	5A Sherman	2A Lafayette*	6A Knox
	4A Johnson	5A Smith	2A Lafourche*	6A Lincoln
4A Allen	4A Kearny	4A Stafford	3A La Salle*	6A Oxford
4A Anderson	4A Kingman	4A Stanton	3A Lincoln*	6A Penobscot
4A Atchison	4A Kiowa	4A Stevens	2A Livingston*	6A Piscataquis
4A Barber	4A Labette	4A Sumner	3A Madison*	6A Sagadahoc
4A Barton	5A Lane	5A Thomas	3A Morehouse	6A Somerset
4A Bourbon	4A Leavenworth	5A Trego	3A Natchitoches*	6A Waldo
4A Brown	4A Lincoln	4A Wabaunsee	2A Orleans*	6A Washington
4A Butler	4A Linn	5A Wallace	3A Ouachita*	6A York
4A Chase	5A Logan	4A Washington	2A Plaquemines*	
4A Chautauqua	4A Lyon	5A Wichita	2A Pointe Coupee*	**MARYLAND**
4A Cherokee	4A Marion	4A Wilson	2A Rapides*	
5A Cheyenne	4A Marshall	4A Woodson	3A Red River*	4A Allegany
4A Clark	4A McPherson	4A Wyandotte	3A Richland*	4A Anne Arundel
4A Clay	4A Meade		3A Sabine*	4A Baltimore
5A Cloud	4A Miami	**KENTUCKY**	2A St. Bernard*	4A Baltimore (city)
4A Coffey	5A Mitchell		2A St. Charles*	4A Calvert
4A Comanche	4A Montgomery	4A (all)	2A St. Helena*	4A Caroline
4A Cowley	4A Morris		2A St. James*	4A Carroll
4A Crawford	4A Morton	**LOUISIANA**	2A St. John the Baptist*	4A Cecil
5A Decatur	4A Nemaha		2A St. Landry*	4A Charles
4A Dickinson	4A Neosho	2A Acadia*	2A St. Martin*	4A Dorchester
4A Doniphan	5A Ness	2A Allen*	2A St. Mary*	4A Frederick
4A Douglas	5A Norton	2A Ascension*	2A St. Tammany*	5A Garrett
4A Edwards	4A Osage	2A Assumption*	2A Tangipahoa*	4A Harford
4A Elk	5A Osborne	2A Avoyelles*	3A Tensas*	4A Howard
5A Ellis	4A Ottawa	2A Beauregard*	2A Terrebonne*	4A Kent
4A Ellsworth	4A Pawnee	3A Bienville*	3A Union*	4A Montgomery
4A Finney	5A Phillips	3A Bossier*	2A Vermilion*	4A Prince George's
4A Ford	4A Pottawatomie	3A Caddo*	3A Vernon*	4A Queen Anne's
4A Franklin	4A Pratt	2A Calcasieu*	2A Washington*	4A Somerset
4A Geary	5A Rawlins	3A Caldwell*	3A Webster*	4A St. Mary's
5A Gove	4A Reno	2A Cameron*	2A West Baton Rouge*	4A Talbot
5A Graham	5A Republic	3A Catahoula*	3A West Carroll	4A Washington
4A Grant	4A Rice	3A Claiborne*	2A West Feliciana*	4A Wicomico
4A Gray	4A Riley	3A Concordia*	3A Winn*	4A Worcester
5A Greeley	5A Rooks	3A De Soto*		
4A Greenwood	4A Rush	2A East Baton Rouge*	**MAINE**	**MASSACHSETTS**
5A Hamilton	4A Russell	3A East Carroll		
4A Harper	4A Saline	2A East Feliciana*	6A Androscoggin	5A (all)
4A Harvey	5A Scott	2A Evangeline*	7 Aroostook	**MICHIGAN**
		3A Franklin*		
		3A Grant*		6A Alcona
		2A Iberia*		6A Alger

(continued)

TABLE C301.1—continued
CLIMATE ZONES, MOISTURE REGIMES, AND WARM-HUMID
DESIGNATIONS BY STATE, COUNTY AND TERRITORY

5A Allegan	7 Mackinac	6A Carver	7 Otter Tail	3A Clarke
6A Alpena	5A Macomb	7 Cass	7 Pennington	3A Clay
6A Antrim	6A Manistee	6A Chippewa	7 Pine	3A Coahoma
6A Arenac	6A Marquette	6A Chisago	6A Pipestone	3A Copiah*
7 Baraga	6A Mason	7 Clay	7 Polk	3A Covington*
5A Barry	6A Mecosta	7 Clearwater	6A Pope	3A DeSoto
5A Bay	6A Menominee	7 Cook	6A Ramsey	3A Forrest*
6A Benzie	5A Midland	6A Cottonwood	7 Red Lake	3A Franklin*
5A Berrien	6A Missaukee	7 Crow Wing	6A Redwood	3A George*
5A Branch	5A Monroe	6A Dakota	6A Renville	3A Greene*
5A Calhoun	5A Montcalm	6A Dodge	6A Rice	3A Grenada
5A Cass	6A Montmorency	6A Douglas	6A Rock	2A Hancock*
6A Charlevoix	5A Muskegon	6A Faribault	7 Roseau	2A Harrison*
6A Cheboygan	6A Newaygo	6A Fillmore	6A Scott	3A Hinds*
7 Chippewa	5A Oakland	6A Freeborn	6A Sherburne	3A Holmes
6A Clare	6A Oceana	6A Goodhue	6A Sibley	3A Humphreys
5A Clinton	6A Ogemaw	7 Grant	6A Stearns	3A Issaquena
6A Crawford	7 Ontonagon	6A Hennepin	6A Steele	3A Itawamba
6A Delta	6A Osceola	6A Houston	6A Stevens	2A Jackson*
6A Dickinson	6A Oscoda	7 Hubbard	7 St. Louis	3A Jasper
5A Eaton	6A Otsego	6A Isanti	6A Swift	3A Jefferson*
6A Emmet	5A Ottawa	7 Itasca	6A Todd	3A Jefferson Davis*
5A Genesee	6A Presque Isle	6A Jackson	6A Traverse	3A Jones*
6A Gladwin	6A Roscommon	7 Kanabec	6A Wabasha	3A Kemper
7 Gogebic	5A Saginaw	6A Kandiyohi	7 Wadena	3A Lafayette
6A Grand Traverse	6A Sanilac	7 Kittson	6A Waseca	3A Lamar*
5A Gratiot	7 Schoolcraft	7 Koochiching	6A Washington	3A Lauderdale
5A Hillsdale	5A Shiawassee	6A Lac qui Parle	6A Watonwan	3A Lawrence*
7 Houghton	5A St. Clair	7 Lake	7 Wilkin	3A Leake
6A Huron	5A St. Joseph	7 Lake of the Woods	6A Winona	3A Lee
5A Ingham	5A Tuscola	6A Le Sueur	6A Wright	3A Leflore
5A Ionia	5A Van Buren	6A Lincoln	6A Yellow	3A Lincoln*
6A Iosco	5A Washtenaw	6A Lyon	Medicine	3A Lowndes
7 Iron	5A Wayne	7 Mahnomen		3A Madison
6A Isabella	6A Wexford	7 Marshall	**MISSISSIPPI**	3A Marion*
5A Jackson		6A Martin		3A Marshall
5A Kalamazoo	**MINNESOTA**	6A McLeod	3A Adams*	3A Monroe
6A Kalkaska		6A Meeker	3A Alcorn	3A Montgomery
5A Kent	7 Aitkin	7 Mille Lacs	3A Amite*	3A Neshoba
7 Keweenaw	6A Anoka	6A Morrison	3A Attala	3A Newton
6A Lake	7 Becker	6A Mower	3A Benton	3A Noxubee
5A Lapeer	7 Beltrami	6A Murray	3A Bolivar	3A Oktibbeha
6A Leelanau	6A Benton	6A Nicollet	3A Calhoun	3A Panola
5A Lenawee	6A Big Stone	6A Nobles	3A Carroll	2A Pearl River*
5A Livingston	6A Blue Earth	7 Norman	3A Chickasaw	3A Perry*
7 Luce	6A Brown	6A Olmsted	3A Choctaw	3A Pike*
	7 Carlton		3A Claiborne*	

(continued)

3A Pontotoc
3A Prentiss
3A Quitman
3A Rankin*
3A Scott
3A Sharkey
3A Simpson*
3A Smith*
2A Stone*
3A Sunflower
3A Tallahatchie
3A Tate
3A Tippah
3A Tishomingo
3A Tunica
3A Union
3A Walthall*
3A Warren*
3A Washington
3A Wayne*
3A Webster
3A Wilkinson*
3A Winston
3A Yalobusha
3A Yazoo

MISSOURI

5A Adair
5A Andrew
5A Atchison
4A Audrain
4A Barry
4A Barton
4A Bates
4A Benton
4A Bollinger
4A Boone
5A Buchanan
4A Butler
5A Caldwell
4A Callaway
4A Camden
4A Cape Girardeau
4A Carroll
4A Carter
4A Cass
4A Cedar

5A Chariton
4A Christian
5A Clark
4A Clay
5A Clinton
4A Cole
4A Cooper
4A Crawford
4A Dade
4A Dallas
5A Daviess
5A DeKalb
4A Dent
4A Douglas
4A Dunklin
4A Franklin
4A Gasconade
5A Gentry
4A Greene
5A Grundy
5A Harrison
4A Henry
4A Hickory
5A Holt
4A Howard
4A Howell
4A Iron
4A Jackson
4A Jasper
4A Jefferson
4A Johnson
5A Knox
4A Laclede
4A Lafayette
4A Lawrence
5A Lewis
4A Lincoln
5A Linn
5A Livingston
5A Macon
4A Madison
4A Maries
5A Marion
4A McDonald
5A Mercer
4A Miller

4A Mississippi
4A Moniteau
4A Monroe
4A Montgomery
4A Morgan
4A New Madrid
4A Newton
5A Nodaway
4A Oregon
4A Osage
4A Ozark
4A Pemiscot
4A Perry
4A Pettis
4A Phelps
5A Pike
4A Platte
4A Polk
4A Pulaski
5A Putnam
5A Ralls
4A Randolph
4A Ray
4A Reynolds
4A Ripley
4A Saline
5A Schuyler
5A Scotland
4A Scott
4A Shannon
5A Shelby
4A St. Charles
4A St. Clair
4A Ste. Genevieve
4A St. Francois
4A St. Louis
4A St. Louis (city)
4A Stoddard
4A Stone
5A Sullivan
4A Taney
4A Texas
4A Vernon
4A Warren
4A Washington
4A Wayne

4A Webster
5A Worth
4A Wright

MONTANA

6B (all)

NEBRASKA

5A (all)

NEVADA

5B Carson City (city)
5B Churchill
3B Clark
5B Douglas
5B Elko
5B Esmeralda
5B Eureka
5B Humboldt
5B Lander
5B Lincoln
5B Lyon
5B Mineral
5B Nye
5B Pershing
5B Storey
5B Washoe
5B White Pine

NEW HAMPSHIRE

6A Belknap
6A Carroll
5A Cheshire
6A Coos
6A Grafton
5A Hillsborough
6A Merrimack
5A Rockingham
5A Strafford
6A Sullivan

NEW JERSEY

4A Atlantic
5A Bergen
4A Burlington
4A Camden
4A Cape May

4A Cumberland
4A Essex
4A Gloucester
4A Hudson
5A Hunterdon
5A Mercer
4A Middlesex
4A Monmouth
5A Morris
4A Ocean
5A Passaic
4A Salem
5A Somerset
5A Sussex
4A Union
5A Warren

NEW MEXICO

4B Bernalillo
5B Catron
3B Chaves
4B Cibola
5B Colfax
4B Curry
4B DeBaca
3B Dona Ana
3B Eddy
4B Grant
4B Guadalupe
5B Harding
3B Hidalgo
3B Lea
4B Lincoln
5B Los Alamos
3B Luna
5B McKinley
5B Mora
3B Otero
4B Quay
5B Rio Arriba
4B Roosevelt
5B Sandoval
5B San Juan
5B San Miguel
5B Santa Fe
4B Sierra
4B Socorro

(continued)

5B Taos
5B Torrance
4B Union
4B Valencia

NEW YORK

5A Albany
6A Allegany
4A Bronx
6A Broome
6A Cattaraugus
5A Cayuga
5A Chautauqua
5A Chemung
6A Chenango
6A Clinton
5A Columbia
5A Cortland
6A Delaware
5A Dutchess
5A Erie
6A Essex
6A Franklin
6A Fulton
5A Genesee
5A Greene
6A Hamilton
6A Herkimer
6A Jefferson
4A Kings
6A Lewis
5A Livingston
6A Madison
5A Monroe
6A Montgomery
4A Nassau
4A New York
5A Niagara
6A Oneida
5A Onondaga
5A Ontario
5A Orange
5A Orleans
5A Oswego
6A Otsego
5A Putnam

4A Queens
5A Rensselaer
4A Richmond
5A Rockland
5A Saratoga
5A Schenectady
6A Schoharie
5A Schuyler
5A Seneca
6A Steuben
6A St. Lawrence
4A Suffolk
6A Sullivan
5A Tioga
6A Tompkins
6A Ulster
6A Warren
5A Washington
5A Wayne
4A Westchester
6A Wyoming
5A Yates

NORTH CAROLINA

4A Alamance
4A Alexander
5A Alleghany
3A Anson
5A Ashe
5A Avery
3A Beaufort
4A Bertie
3A Bladen
3A Brunswick*
4A Buncombe
4A Burke
3A Cabarrus
4A Caldwell
3A Camden
3A Carteret*
4A Caswell
4A Catawba
4A Chatham
4A Cherokee
3A Chowan

4A Clay
4A Cleveland
3A Columbus*
3A Craven
3A Cumberland
3A Currituck
3A Dare
3A Davidson
4A Davie
3A Duplin
4A Durham
3A Edgecombe
4A Forsyth
4A Franklin
3A Gaston
4A Gates
4A Graham
4A Granville
3A Greene
4A Guilford
4A Halifax
4A Harnett
4A Haywood
4A Henderson
4A Hertford
3A Hoke
3A Hyde
4A Iredell
4A Jackson
3A Johnston
3A Jones
4A Lee
3A Lenoir
4A Lincoln
4A Macon
4A Madison
3A Martin
4A McDowell
3A Mecklenburg
5A Mitchell
3A Montgomery
3A Moore
4A Nash
3A New Hanover*
4A Northampton
3A Onslow*

4A Orange
3A Pamlico
3A Pasquotank
3A Pender*
3A Perquimans
4A Person
3A Pitt
4A Polk
3A Randolph
3A Richmond
3A Robeson
4A Rockingham
3A Rowan
4A Rutherford
3A Sampson
3A Scotland
3A Stanly
4A Stokes
4A Surry
4A Swain
4A Transylvania
3A Tyrrell
3A Union
4A Vance
4A Wake
4A Warren
3A Washington
5A Watauga
3A Wayne
4A Wilkes
3A Wilson
4A Yadkin
5A Yancey

NORTH DAKOTA

6A Adams
7 Barnes
7 Benson
6A Billings
7 Bottineau
6A Bowman
7 Burke
6A Burleigh
7 Cass
7 Cavalier
6A Dickey

7 Divide
6A Dunn
7 Eddy
6A Emmons
7 Foster
6A Golden Valley
7 Grand Forks
6A Grant
7 Griggs
6A Hettinger
7 Kidder
6A LaMoure
6A Logan
7 McHenry
6A McIntosh
6A McKenzie
7 McLean
6A Mercer
6A Morton
7 Mountrail
7 Nelson
6A Oliver
7 Pembina
7 Pierce
7 Ramsey
6A Ransom
7 Renville
6A Richland
7 Rolette
6A Sargent
7 Sheridan
6A Sioux
6A Slope
6A Stark
7 Steele
7 Stutsman
7 Towner
7 Traill
7 Walsh
7 Ward
7 Wells
7 Williams

OHIO

4A Adams
5A Allen

(continued)

TABLE C301.1—continued
**CLIMATE ZONES, MOISTURE REGIMES, AND WARM-HUMID
DESIGNATIONS BY STATE, COUNTY AND TERRITORY**

5A Ashland
5A Ashtabula
5A Athens
5A Auglaize
5A Belmont
4A Brown
5A Butler
5A Carroll
5A Champaign
5A Clark
4A Clermont
5A Clinton
5A Columbiana
5A Coshocton
5A Crawford
5A Cuyahoga
5A Darke
5A Defiance
5A Delaware
5A Erie
5A Fairfield
5A Fayette
5A Franklin
5A Fulton
4A Gallia
5A Geauga
5A Greene
5A Guernsey
4A Hamilton
5A Hancock
5A Hardin
5A Harrison
5A Henry
5A Highland
5A Hocking
5A Holmes
5A Huron
5A Jackson
5A Jefferson
5A Knox
5A Lake
4A Lawrence
5A Licking
5A Logan
5A Lorain
5A Lucas
5A Madison

5A Mahoning
5A Marion
5A Medina
5A Meigs
5A Mercer
5A Miami
5A Monroe
5A Montgomery
5A Morgan
5A Morrow
5A Muskingum
5A Noble
5A Ottawa
5A Paulding
5A Perry
5A Pickaway
4A Pike
5A Portage
5A Preble
5A Putnam
5A Richland
5A Ross
5A Sandusky
4A Scioto
5A Seneca
5A Shelby
5A Stark
5A Summit
5A Trumbull
5A Tuscarawas
5A Union
5A Van Wert
5A Vinton
5A Warren
4A Washington
5A Wayne
5A Williams
5A Wood
5A Wyandot

OKLAHOMA

3A Adair
3A Alfalfa
3A Atoka
4B Beaver
3A Beckham
3A Blaine

3A Bryan
3A Caddo
3A Canadian
3A Carter
3A Cherokee
3A Choctaw
4B Cimarron
3A Cleveland
3A Coal
3A Comanche
3A Cotton
3A Craig
3A Creek
3A Custer
3A Delaware
3A Dewey
3A Ellis
3A Garfield
3A Garvin
3A Grady
3A Grant
3A Greer
3A Harmon
3A Harper
3A Haskell
3A Hughes
3A Jackson
3A Jefferson
3A Johnston
3A Kay
3A Kingfisher
3A Kiowa
3A Latimer
3A Le Flore
3A Lincoln
3A Logan
3A Love
3A Major
3A Marshall
3A Mayes
3A McClain
3A McCurtain
3A McIntosh
3A Murray
3A Muskogee
3A Noble
3A Nowata

3A Okfuskee
3A Oklahoma
3A Okmulgee
3A Osage
3A Ottawa
3A Pawnee
3A Payne
3A Pittsburg
3A Pontotoc
3A Pottawatomie
3A Pushmataha
3A Roger Mills
3A Rogers
3A Seminole
3A Sequoyah
3A Stephens
4B Texas
3A Tillman
3A Tulsa
3A Wagoner
3A Washington
3A Washita
3A Woods
3A Woodward

OREGON

5B Baker
4C Benton
4C Clackamas
4C Clatsop
4C Columbia
4C Coos
5B Crook
4C Curry
5B Deschutes
4C Douglas
5B Gilliam
5B Grant
5B Harney
5B Hood River
4C Jackson
5B Jefferson
4C Josephine
5B Klamath
5B Lake
4C Lane
4C Lincoln

4C Linn
5B Malheur
4C Marion
5B Morrow
4C Multnomah
4C Polk
5B Sherman
4C Tillamook
5B Umatilla
5B Union
5B Wallowa
5B Wasco
4C Washington
5B Wheeler
4C Yamhill

PENNSYLVANIA

5A Adams
5A Allegheny
5A Armstrong
5A Beaver
5A Bedford
5A Berks
5A Blair
5A Bradford
4A Bucks
5A Butler
5A Cambria
6A Cameron
5A Carbon
5A Centre
4A Chester
5A Clarion
6A Clearfield
5A Clinton
5A Columbia
5A Crawford
5A Cumberland
5A Dauphin
4A Delaware
6A Elk
5A Erie
5A Fayette
5A Forest
5A Franklin
5A Fulton
5A Greene

(continued)

**TABLE C301.1—continued
CLIMATE ZONES, MOISTURE REGIMES, AND WARM-HUMID
DESIGNATIONS BY STATE, COUNTY AND TERRITORY**

5A Huntingdon
5A Indiana
5A Jefferson
5A Juniata
5A Lackawanna
5A Lancaster
5A Lawrence
5A Lebanon
5A Lehigh
5A Luzerne
5A Lycoming
6A McKean
5A Mercer
5A Mifflin
5A Monroe
4A Montgomery
5A Montour
5A Northampton
5A Northumberland
5A Perry
4A Philadelphia
5A Pike
6A Potter
5A Schuylkill
5A Snyder
5A Somerset
5A Sullivan
6A Susquehanna
6A Tioga
5A Union
5A Venango
5A Warren
5A Washington
6A Wayne
5A Westmoreland
5A Wyoming
4A York

RHODE ISLAND

5A (all)

**SOUTH
CAROLINA**

3A Abbeville
3A Aiken
3A Allendale*
3A Anderson

3A Bamberg*
3A Barnwell*
3A Beaufort*
3A Berkeley*
3A Calhoun
3A Charleston*
3A Cherokee
3A Chester
3A Chesterfield
3A Clarendon
3A Colleton*
3A Darlington
3A Dillon
3A Dorchester*
3A Edgefield
3A Fairfield
3A Florence
3A Georgetown*
3A Greenville
3A Greenwood
3A Hampton*
3A Horry*
3A Jasper*
3A Kershaw
3A Lancaster
3A Laurens
3A Lee
3A Lexington
3A Marion
3A Marlboro
3A McCormick
3A Newberry
3A Oconee
3A Orangeburg
3A Pickens
3A Richland
3A Saluda
3A Spartanburg
3A Sumter
3A Union
3A Williamsburg
3A York

SOUTH DAKOTA

6A Aurora
6A Beadle

5A Bennett
5A Bon Homme
6A Brookings
6A Brown
6A Brule
6A Buffalo
6A Butte
6A Campbell
5A Charles Mix
6A Clark
5A Clay
6A Codington
6A Corson
6A Custer
6A Davison
6A Day
6A Deuel
6A Dewey
5A Douglas
6A Edmunds
6A Fall River
6A Faulk
6A Grant
5A Gregory
6A Haakon
6A Hamlin
6A Hand
6A Hanson
6A Harding
6A Hughes
5A Hutchinson
6A Hyde
5A Jackson
6A Jerauld
6A Jones
6A Kingsbury
6A Lake
6A Lawrence
6A Lincoln
6A Lyman
6A Marshall
6A McCook
6A McPherson
6A Meade
5A Mellette
6A Miner

6A Minnehaha
6A Moody
6A Pennington
6A Perkins
6A Potter
6A Roberts
6A Sanborn
6A Shannon
6A Spink
6A Stanley
6A Sully
5A Todd
5A Tripp
6A Turner
5A Union
6A Walworth
5A Yankton
6A Ziebach

TENNESSEE

4A Anderson
4A Bedford
4A Benton
4A Bledsoe
4A Blount
4A Bradley
4A Campbell
4A Cannon
4A Carroll
4A Carter
4A Cheatham
3A Chester
4A Claiborne
4A Clay
4A Cocke
4A Coffee
3A Crockett
4A Cumberland
4A Davidson
4A Decatur
4A DeKalb
4A Dickson
3A Dyer
3A Fayette
4A Fentress
4A Franklin

4A Gibson
4A Giles
4A Grainger
4A Greene
4A Grundy
4A Hamblen
4A Hamilton
4A Hancock
3A Hardeman
3A Hardin
4A Hawkins
3A Haywood
3A Henderson
4A Henry
4A Hickman
4A Houston
4A Humphreys
4A Jackson
4A Jefferson
4A Johnson
4A Knox
3A Lake
3A Lauderdale
4A Lawrence
4A Lewis
4A Lincoln
4A Loudon
4A Macon
3A Madison
4A Marion
4A Marshall
4A Maury
4A McMinn
3A McNairy
4A Meigs
4A Monroe
4A Montgomery
4A Moore
4A Morgan
4A Obion
4A Overton
4A Perry
4A Pickett
4A Polk
4A Putnam
4A Rhea

(continued)

TABLE C301.1—continued
CLIMATE ZONES, MOISTURE REGIMES, AND WARM-HUMID
DESIGNATIONS BY STATE, COUNTY AND TERRITORY

4A Roane	3B Brewster	3B Ector	3B Howard	3B McCulloch
4A Robertson	4B Briscoe	2B Edwards*	3B Hudspeth	2A McLennan*
4A Rutherford	2A Brooks*	3A Ellis*	3A Hunt*	2A McMullen*
4A Scott	3A Brown*	3B El Paso	4B Hutchinson	2B Medina*
4A Sequatchie	2A Burleson*	3A Erath*	3B Irion	3B Menard
4A Sevier	3A Burnet*	2A Falls*	3A Jack	3B Midland
3A Shelby	2A Caldwell*	3A Fannin	2A Jackson*	2A Milam*
4A Smith	2A Calhoun*	2A Fayette*	2A Jasper*	3A Mills*
4A Stewart	3B Callahan	3B Fisher	3B Jeff Davis	3B Mitchell
4A Sullivan	2A Cameron*	4B Floyd	2A Jefferson*	3A Montague
4A Sumner	3A Camp*	3B Foard	2A Jim Hogg*	2A Montgomery*
3A Tipton	4B Carson	2A Fort Bend*	2A Jim Wells*	4B Moore
4A Trousdale	3A Cass*	3A Franklin*	3A Johnson*	3A Morris*
4A Unicoi	4B Castro	2A Freestone*	3B Jones	3B Motley
4A Union	2A Chambers*	2B Frio*	2A Karnes*	3A Nacogdoches*
4A Van Buren	2A Cherokee*	3B Gaines	3A Kaufman*	3A Navarro*
4A Warren	3B Childress	2A Galveston*	3A Kendall*	2A Newton*
4A Washington	3A Clay	3B Garza	2A Kenedy*	3B Nolan
4A Wayne	4B Cochran	3A Gillespie*	3B Kent	2A Nueces*
4A Weakley	3B Coke	3B Glasscock	3B Kerr	4B Ochiltree
4A White	3B Coleman	2A Goliad*	3B Kimble	4B Oldham
4A Williamson	3A Collin*	2A Gonzales*	3B King	2A Orange*
4A Wilson	3B Collingsworth	4B Gray	2B Kinney*	3A Palo Pinto*
	2A Colorado*	3A Grayson	2A Kleberg*	3A Panola*
TEXAS	2A Comal*	3A Gregg*	3B Knox	3A Parker*
	3A Comanche*	2A Grimes*	3A Lamar*	4B Parmer
2A Anderson*	3B Concho	2A Guadalupe*	4B Lamb	3B Pecos
3B Andrews	3A Cooke	4B Hale	3A Lampasas*	2A Polk*
2A Angelina*	2A Coryell*	3B Hall	2B La Salle*	4B Potter
2A Aransas*	3B Cottle	3A Hamilton*	2A Lavaca*	3B Presidio
3A Archer	3B Crane	4B Hansford	2A Lee*	3A Rains*
4B Armstrong	3B Crockett	3B Hardeman	2A Leon*	4B Randall
2A Atascosa*	3B Crosby	2A Hardin*	2A Liberty*	3B Reagan
2A Austin*	3B Culberson	2A Harris*	2A Limestone*	2B Real*
4B Bailey	4B Dallam	3A Harrison*	4B Lipscomb	3A Red River*
2B Bandera*	3A Dallas*	4B Hartley	2A Live Oak*	3B Reeves
2A Bastrop*	3B Dawson	3B Haskell	3A Llano*	2A Refugio*
3B Baylor	4B Deaf Smith	2A Hays*	3B Loving	4B Roberts
2A Bee*	3A Delta	3B Hemphill	3B Lubbock	2A Robertson*
2A Bell*	3A Denton*	3A Henderson*	3B Lynn	3A Rockwall*
2A Bexar*	2A DeWitt*	2A Hidalgo*	2A Madison*	3B Runnels
3A Blanco*	3B Dickens	2A Hill*	3A Marion*	3A Rusk*
3B Borden	2B Dimmit*	4B Hockley	3B Martin	3A Sabine*
2A Bosque*	4B Donley	3A Hood*	3B Mason	3A San Augustine*
3A Bowie*	2A Duval*	3A Hopkins*	2A Matagorda*	2A San Jacinto*
2A Brazoria*	3A Eastland	2A Houston*	2B Maverick*	2A San Patricio*
2A Brazos*				

(continued)

TABLE C301.1—continued
CLIMATE ZONES, MOISTURE REGIMES, AND WARM-HUMID
DESIGNATIONS BY STATE, COUNTY AND TERRITORY

3A San Saba*
3B Schleicher
3B Scurry
3B Shackelford
3A Shelby*
4B Sherman
3A Smith*
3A Somervell*
2A Starr*
3A Stephens
3B Sterling
3B Stonewall
3B Sutton
4B Swisher
3A Tarrant*
3B Taylor
3B Terrell
3B Terry
3B Throckmorton
3A Titus*
3B Tom Green
2A Travis*
2A Trinity*
2A Tyler*
3A Upshur*
3B Upton
2B Uvalde*
2B Val Verde*
3A Van Zandt*
2A Victoria*
2A Walker*
2A Waller*
3B Ward
2A Washington*
2B Webb*
2A Wharton*
3B Wheeler
3A Wichita
3B Wilbarger
2A Willacy*
2A Williamson*
2A Wilson*
3B Winkler
3A Wise
3A Wood*
4B Yoakum

3A Young
2B Zapata*
2B Zavala*

UTAH

5B Beaver
6B Box Elder
6B Cache
6B Carbon
6B Daggett
5B Davis
6B Duchesne
5B Emery
5B Garfield
5B Grand
5B Iron
5B Juab
5B Kane
5B Millard
6B Morgan
5B Piute
6B Rich
5B Salt Lake
5B San Juan
5B Sanpete
5B Sevier
6B Summit
5B Tooele
6B Uintah
5B Utah
6B Wasatch
3B Washington
5B Wayne
5B Weber

VERMONT

6A (all)

VIRGINIA

4A (all)

WASHINGTON

5B Adams
5B Asotin
5B Benton
5B Chelan
4C Clallam

4C Clark
5B Columbia
4C Cowlitz
5B Douglas
6B Ferry
5B Franklin
5B Garfield
5B Grant
4C Grays Harbor
4C Island
4C Jefferson
4C King
4C Kitsap
5B Kittitas
5B Klickitat
4C Lewis
5B Lincoln
4C Mason
6B Okanogan
4C Pacific
6B Pend Oreille
4C Pierce
4C San Juan
4C Skagit
5B Skamania
4C Snohomish
5B Spokane
6B Stevens
4C Thurston
4C Wahkiakum
5B Walla Walla
4C Whatcom
5B Whitman
5B Yakima

WEST VIRGINIA

5A Barbour
4A Berkeley
4A Boone
4A Braxton
5A Brooke
4A Cabell
4A Calhoun
4A Clay
5A Doddridge
5A Fayette

4A Gilmer
5A Grant
5A Greenbrier
5A Hampshire
5A Hancock
5A Hardy
5A Harrison
4A Jackson
4A Jefferson
4A Kanawha
5A Lewis
4A Lincoln
4A Logan
5A Marion
5A Marshall
4A Mason
4A McDowell
4A Mercer
5A Mineral
4A Mingo
5A Monongalia
4A Monroe
4A Morgan
5A Nicholas
5A Ohio
5A Pendleton
4A Pleasants
5A Pocahontas
5A Preston
4A Putnam
5A Raleigh
5A Randolph
4A Ritchie
4A Roane
5A Summers
5A Taylor
5A Tucker
4A Tyler
5A Upshur
4A Wayne
5A Webster
5A Wetzel
4A Wirt
5A Wood
4A Wyoming

WISCONSIN

6A Adams
7 Ashland
6A Barron
7 Bayfield
6A Brown
6A Buffalo
7 Burnett
6A Calumet
6A Chippewa
6A Clark
6A Columbia
6A Crawford
6A Dane
6A Dodge
6A Door
7 Douglas
6A Dunn
6A Eau Claire
7 Florence
6A Fond du Lac
7 Forest
6A Grant
6A Green
6A Green Lake
6A Iowa
7 Iron
6A Jackson
6A Jefferson
6A Juneau
6A Kenosha
6A Kewaunee
6A La Crosse
6A Lafayette
7 Langlade
7 Lincoln
6A Manitowoc
6A Marathon
6A Marinette
6A Marquette
6A Menominee
6A Milwaukee
6A Monroe
6A Oconto
7 Oneida
6A Outagamie

(continued)

TABLE C301.1—continued
CLIMATE ZONES, MOISTURE REGIMES, AND WARM-HUMID
DESIGNATIONS BY STATE, COUNTY AND TERRITORY

6A Ozaukee	7 Taylor	6B Big Horn	6B Sheridan	**NORTHERN MARIANA ISLANDS**
6A Pepin	6A Trempealeau	6B Campbell	7 Sublette	
6A Pierce	6A Vernon	6B Carbon	6B Sweetwater	1A (all)*
6A Polk	7 Vilas	6B Converse	7 Teton	**PUERTO RICO**
6A Portage	6A Walworth	6B Crook	6B Uinta	
7 Price	7 Washburn	6B Fremont	6B Washakie	1A (all)*
6A Racine	6A Washington	5B Goshen	6B Weston	**VIRGIN ISLANDS**
6A Richland	6A Waukesha	6B Hot Springs	**US TERRITORIES**	1A (all)*
6A Rock	6A Waupaca	6B Johnson		
6A Rusk	6A Waushara	6B Laramie	**AMERICAN SAMOA**	
6A Sauk	6A Winnebago	7 Lincoln		
7 Sawyer	6A Wood	6B Natrona	1A (all)*	
6A Shawano		6B Niobrara	**GUAM**	
6A Sheboygan	**WYOMING**	6B Park		
6A St. Croix	6B Albany	5B Platte	1A (all)*	

TABLE C301.3(1)
INTERNATIONAL CLIMATE ZONE DEFINITIONS

MAJOR CLIMATE TYPE DEFINITIONS
Marine (C) Definition—Locations meeting all four criteria: 1. Mean temperature of coldest month between -3°C (27°F) and 18°C (65°F). 2. Warmest month mean < 22°C (72°F). 3. At least four months with mean temperatures over 10°C (50°F). 4. Dry season in summer. The month with the heaviest precipitation in the cold season has at least three times as much precipitation as the month with the least precipitation in the rest of the year. The cold season is October through March in the Northern Hemisphere and April through September in the Southern Hemisphere.
Dry (B) Definition—Locations meeting the following criteria: Not marine and $P_{in} < 0.44 \times (TF - 19.5)$ [$P_{cm} < 2.0 \times (TC + 7)$ in SI units] where: P_{in} = Annual precipitation in inches (cm) T = Annual mean temperature in °F (°C)
Moist (A) Definition—Locations that are not marine and not dry.
Warm-humid Definition—Moist (A) locations where either of the following wet-bulb temperature conditions shall occur during the warmest six consecutive months of the year: 1. 67°F (19.4°C) or higher for 3,000 or more hours; or 2. 73°F (22.8°C) or higher for 1,500 or more hours.

For SI: °C = [(°F)-32]/1.8, 1 inch = 2.54 cm.

TABLE C301.3(2)
INTERNATIONAL CLIMATE ZONE DEFINITIONS

ZONE NUMBER	THERMAL CRITERIA	
	IP Units	SI Units
1	$9000 < \text{CDD}50°\text{F}$	$5000 < \text{CDD}10°\text{C}$
2	$6300 < \text{CDD}50°\text{F} \leq 9000$	$3500 < \text{CDD}10°\text{C} \leq 5000$
3A and 3B	$4500 < \text{CDD}50°\text{F} \leq 6300$ AND $\text{HDD}65°\text{F} \leq 5400$	$2500 < \text{CDD}10°\text{C} \leq 3500$ AND $\text{HDD}18°\text{C} \leq 3000$
4A and 4B	$\text{CDD}50°\text{F} \leq 4500$ AND $\text{HDD}65°\text{F} \leq 5400$	$\text{CDD}10°\text{C} \leq 2500$ AND $\text{HDD}18°\text{C} \leq 3000$
3C	$\text{HDD}65°\text{F} \leq 3600$	$\text{HDD}18°\text{C} \leq 2000$
4C	$3600 < \text{HDD}65°\text{F} \leq 5400$	$2000 < \text{HDD}18°\text{C} \leq 3000$
5	$5400 < \text{HDD}65°\text{F} \leq 7200$	$3000 < \text{HDD}18°\text{C} \leq 4000$
6	$7200 < \text{HDD}65°\text{F} \leq 9000$	$4000 < \text{HDD}18°\text{C} \leq 5000$
7	$9000 < \text{HDD}65°\text{F} \leq 12600$	$5000 < \text{HDD}18°\text{C} \leq 7000$
8	$12600 < \text{HDD}65°\text{F}$	$7000 < \text{HDD}18°\text{C}$

For SI: $°\text{C} = [(°\text{F})-32]/1.8$.

SECTION C302
DESIGN CONDITIONS

C302.1 Interior design conditions. The interior design temperatures used for heating and cooling load calculations shall be a maximum of 72°F (22°C) for heating and minimum of 75°F (24°C) for cooling.

SECTION C303
MATERIALS, SYSTEMS AND EQUIPMENT

C303.1 Identification. Materials, systems and equipment shall be identified in a manner that will allow a determination of compliance with the applicable provisions of this code.

C303.1.1 Building thermal envelope insulation. An *R*-value identification mark shall be applied by the manufacturer to each piece of *building thermal envelope* insulation 12 inches (305 mm) or greater in width. Alternately, the insulation installers shall provide a certification listing the type, manufacturer and *R*-value of insulation installed in each element of the *building thermal envelope*. For blown or sprayed insulation (fiberglass and cellulose), the initial installed thickness, settled thickness, settled *R*-value, installed density, coverage area and number of bags installed shall be *listed* on the certification. For sprayed polyurethane foam (SPF) insulation, the installed thickness of the areas covered and *R*-value of installed thickness shall be *listed* on the certification. The insulation installer shall sign, date and post the certification in a conspicuous location on the job site.

C303.1.1.1 Blown or sprayed roof/ceiling insulation. The thickness of blown-in or sprayed roof/ceiling insulation (fiberglass or cellulose) shall be written in inches (mm) on markers that are installed at least one for every 300 square feet (28 m²) throughout the attic space. The markers shall be affixed to the trusses or joists and marked with the minimum initial installed thickness with numbers a minimum of 1 inch (25 mm) in height. Each marker shall face the attic access opening. Spray polyurethane foam thickness and installed *R*-value shall

be *listed* on certification provided by the insulation installer.

C303.1.2 Insulation mark installation. Insulating materials shall be installed such that the manufacturer's *R*-value mark is readily observable upon inspection.

C303.1.3 Fenestration product rating. *U*-factors of fenestration products (windows, doors and skylights) shall be determined in accordance with NFRC 100 by an accredited, independent laboratory, and labeled and certified by the manufacturer. Products lacking such a labeled *U*-factor shall be assigned a default *U*-factor from Table C303.1.3(1) or C303.1.3(2). The solar heat gain coefficient (SHGC) and *visible transmittance* (VT) of glazed fenestration products (windows, glazed doors and skylights) shall be determined in accordance with NFRC 200 by an accredited, independent laboratory, and labeled and certified by the manufacturer. Products lacking such a labeled SHGC or VT shall be assigned a default SHGC or VT from Table C303.1.3(3).

TABLE C303.1.3(1)
DEFAULT GLAZED FENESTRATION *U*-FACTOR

FRAME TYPE	SINGLE PANE	DOUBLE PANE	SKYLIGHT	
			Single	Double
Metal	1.20	0.80	2.00	1.30
Metal with Thermal Break	1.10	0.65	1.90	1.10
Nonmetal or Metal Clad	0.95	0.55	1.75	1.05
Glazed Block	0.60			

TABLE C303.1.3(2)
DEFAULT DOOR *U*-FACTORS

DOOR TYPE	*U*-FACTOR
Uninsulated Metal	1.20
Insulated Metal	0.60
Wood	0.50
Insulated, nonmetal edge, max 45% glazing, any glazing double pane	0.35

TABLE C303.1.3(3)
DEFAULT GLAZED FENESTRATION SHGC AND VT

	SINGLE GLAZED		DOUBLE GLAZED		GLAZED BLOCK
	Clear	Tinted	Clear	Tinted	
SHGC	0.8	0.7	0.7	0.6	0.6
VT	0.6	0.3	0.6	0.3	0.6

C303.1.4 Insulation product rating. The thermal resistance (*R*-value) of insulation shall be determined in accordance with the U.S. Federal Trade Commission *R*-value rule (CFR Title 16, Part 460) in units of h × ft^2 × °F/Btu at a mean temperature of 75°F (24°C).

C303.2 Installation. All materials, systems and equipment shall be installed in accordance with the manufacturer's installation instructions and the *International Building Code*.

C303.2.1 Protection of exposed foundation insulation. Insulation applied to the exterior of basement walls, crawlspace walls and the perimeter of slab-on-grade floors shall have a rigid, opaque and weather-resistant protective covering to prevent the degradation of the insulation's thermal performance. The protective covering shall cover the exposed exterior insulation and extend a minimum of 6 inches (153 mm) below grade.

C303.3 Maintenance information. Maintenance instructions shall be furnished for equipment and systems that require preventive maintenance. Required regular maintenance actions shall be clearly stated and incorporated on a readily accessible label. The label shall include the title or publication number for the operation and maintenance manual for that particular model and type of product.

CHAPTER 4 [CE]

COMMERCIAL ENERGY EFFICIENCY

SECTION C401
GENERAL

C401.1 Scope. The requirements contained in this chapter are applicable to commercial buildings, or portions of commercial buildings.

C401.2 Application. Commercial buildings shall comply with one of the following:

1. The requirements of ANSI/ASHRAE/IESNA 90.1.

2. The requirements of Sections C402, C403, C404 and C405. In addition, commercial buildings shall comply with either Section C406.2, C406.3 or C406.4.

3. The requirements of Section C407, C402.4, C403.2, C404, C405.2, C405.3, C405.4, C405.6 and C405.7. The building energy cost shall be equal to or less than 85 percent of the standard reference design building.

C401.2.1 Application to existing buildings. Additions, alterations and repairs to existing buildings shall comply with one of the following:

1. Sections C402, C403, C404 and C405; or

2. ANSI/ASHRAE/IESNA 90.1.

SECTION C402
BUILDING ENVELOPE REQUIREMENTS

C402.1 General (Prescriptive). The building thermal envelope shall comply with Section C402.1.1. Section C402.1.2 shall be permitted as an alternative to the *R*-values specified in Section C402.1.1.

C402.1.1 Insulation and fenestration criteria. The *building thermal envelope* shall meet the requirements of Tables C402.2 and C402.3 based on the climate zone specified in Chapter 3. Commercial buildings or portions of commercial buildings enclosing Group R occupancies shall use the *R*-values from the "Group R" column of Table C402.2. Commercial buildings or portions of commercial buildings enclosing occupancies other than Group R shall use the *R*-values from the "All other" column of Table C402.2. Buildings with a vertical fenestration area or skylight area that exceeds that allowed in Table C402.3 shall comply with the building envelope provisions of ANSI/ASHRAE/IESNA 90.1.

C402.1.2 *U*-factor alternative. An assembly with a *U*-factor, *C*-factor, or *F*-factor equal or less than that specified in Table C402.1.2 shall be permitted as an alternative to the *R*-value in Table C402.2. Commercial buildings or portions of commercial buildings enclosing Group R occupancies shall use the *U*-factor, *C*-factor, or *F*-factor from the "Group R" column of Table C402.1.2. Commercial buildings or portions of commercial buildings enclosing occupancies other than Group R shall use the *U*-factor, *C*-factor or *F*-factor from the "All other" column of Table C402.1.2.

C402.2 Specific insulation requirements (Prescriptive). Opaque assemblies shall comply with Table C402.2. Where two or more layers of continuous insulation board are used in a construction assembly, the continuous insulation boards shall be installed in accordance with Section C303.2. If the continuous insulation board manufacturer's installation instructions do not address installation of two or more layers, the edge joints between each layer of continuous insulation boards shall be staggered.

C402.2.1 Roof assembly. The minimum thermal resistance (*R*-value) of the insulating material installed either between the roof framing or continuously on the roof assembly shall be as specified in Table C402.2, based on construction materials used in the roof assembly. Skylight curbs shall be insulated to the level of roofs with insulation entirely above deck or R-5, whichever is less.

Exceptions:

1. Continuously insulated roof assemblies where the thickness of insulation varies 1 inch (25 mm) or less and where the area-weighted *U*-factor is equivalent to the same assembly with the *R*-value specified in Table C402.2.

2. Unit skylight curbs included as a component of an NFRC 100 rated assembly shall not be required to be insulated.

Insulation installed on a suspended ceiling with removable ceiling tiles shall not be considered part of the minimum thermal resistance of the roof insulation.

C402.2.1.1 Roof solar reflectance and thermal emittance. Low-sloped roofs, with a slope less than 2 units vertical in 12 horizontal, directly above cooled *conditioned spaces* in Climate Zones 1, 2, and 3 shall comply with one or more of the options in Table C402.2.1.1.

Exceptions: The following roofs and portions of roofs are exempt from the requirements in Table C402.2.1.1:

1. Portions of roofs that include or are covered by:

    1.1. Photovoltaic systems or components.

    1.2. Solar air or water heating systems or components.

    1.3. Roof gardens or landscaped roofs.

    1.4. Above-roof decks or walkways.

    1.5. Skylights.

    1.6. HVAC systems, components, and other opaque objects mounted above the roof.

2. Portions of roofs shaded during the peak sun angle on the summer solstice by permanent features of the building, or by permanent features of adjacent buildings.

3. Portions of roofs that are ballasted with a minimum stone ballast of 17 pounds per square foot (psf) (74 kg/m²) or 23 psf (117 kg/m²) pavers.

4. Roofs where a minimum of 75 percent of the roof area meets a minimum of one of the exceptions above.

TABLE C402.2.1.1
MINIMUM ROOF REFLECTANCE AND EMITTANCE OPTIONS[a]

Three-year aged solar reflectance[b] of 0.55 and three-year aged thermal emittance[c] of 0.75
Initial solar reflectance[b] of 0.70 and initial thermal emittance[c] of 0.75
Three-year-aged solar reflectance index[d] of 64
Initial solar reflectance index[d] of 82

a. The use of area-weighted averages to meet these requirements shall be permitted. Materials lacking initial tested values for either solar reflectance or thermal emittance, shall be assigned both an initial solar reflectance of 0.10 and an initial thermal emittance of 0.90. Materials lacking three-year aged tested values for either solar reflectance or thermal emittance shall be assigned both a three-year aged solar reflectance of 0.10 and a three-year aged thermal emittance of 0.90.
b. Solar reflectance tested in accordance with ASTM C 1549, ASTM E 903 or ASTM E 1918.
c. Thermal emittance tested in accordance with ASTM C 1371 or ASTM E 408.
d. Solar reflectance index (SRI) shall be determined in accordance with ASTM E 1980 using a convection coefficient of 2.1 Btu/h × ft² ×°F (12W/m² × K). Calculation of aged SRI shall be based on aged tested values of solar reflectance and thermal emittance. Calculation of initial SRI shall be based on initial tested values of solar reflectance and thermal emittance.

C402.2.2 Classification of walls. Walls associated with the building envelope shall be classified in accordance with Section C402.2.2.1 or C402.2.2.2.

C402.2.2.1 Above-grade walls. Above-grade walls are those walls covered by Section C402.2.3 on the exterior of the building and completely above grade or walls that are more than 15 percent above grade.

C402.2.2.2 Below-grade walls. Below-grade walls covered by Section C402.2.4 are basement or first-story walls associated with the exterior of the building that are at least 85 percent below grade.

C402.2.3 Thermal resistance of above-grade walls. The minimum thermal resistance (R-value) of the insulating materials installed in the wall cavity between the framing members and continuously on the walls shall be as specified in Table C402.2, based on framing type and construction materials used in the wall assembly. The R-value of integral insulation installed in concrete masonry units (CMU) shall not be used in determining compliance with Table C402.2.

"Mass walls" shall include walls weighing not less than:

1. 35 psf (170 kg/m²) of wall surface area; or

2. 25 psf (120 kg/m²) of wall surface area if the material weight is not more than 120 pounds per cubic foot (pcf) (1900 kg/m³).

C402.2.4 Thermal resistance of below-grade walls. The minimum thermal resistance (R-value) of the insulating material installed in, or continuously on, the below-grade walls shall be as specified in Table C402.2, and shall extend to a depth of 10 feet (3048 mm) below the outside finished ground level, or to the level of the floor, whichever is less.

C402.2.5 Floors over outdoor air or unconditioned space. The minimum thermal resistance (R-value) of the insulating material installed either between the floor framing or continuously on the floor assembly shall be as specified in Table C402.2, based on construction materials used in the floor assembly.

"Mass floors" shall include floors weighing not less than:

1. 35 psf (170 kg/m²) of floor surface area; or

2. 25 psf (120 kg/m²) of floor surface area if the material weight is not more than 120 pcf (1,900 kg/m³).

C402.2.6 Slabs on grade. Where the slab on grade is in contact with the ground, the minimum thermal resistance (R-value) of the insulation around the perimeter of unheated or heated slab-on-grade floors shall be as specified in Table C402.2. The insulation shall be placed on the outside of the foundation or on the inside of the foundation wall. The insulation shall extend downward from the top of the slab for a minimum distance as shown in the table or to the top of the footing, whichever is less, or downward to at least the bottom of the slab and then horizontally to the interior or exterior for the total distance shown in the table. Insulation extending away from the building shall be protected by pavement or by a minimum of 10 inches (254 mm) of soil.

Exception: Where the slab-on-grade floor is greater than 24 inches (61 mm) below the finished exterior grade, perimeter insulation is not required.

C402.2.7 Opaque doors. Opaque doors (doors having less than 50 percent glass area) shall meet the applicable requirements for doors as specified in Table C402.2 and be considered as part of the gross area of above-grade walls that are part of the building envelope.

C402.2.8 Insulation of radiant heating systems. Radiant panels, and associated U-bends and headers, designed for sensible heating of an indoor space through heat transfer from the thermally effective panel surfaces to the occupants or indoor space by thermal radiation and natural convection and the bottom surfaces of floor structures incorporating radiant heating shall be insulated with a minimum of R-3.5 (0.62 m²/K × W).

TABLE C402.1.2
OPAQUE THERMAL ENVELOPE ASSEMBLY REQUIREMENTS[a]

CLIMATE ZONE	1		2		3		4 EXCEPT MARINE		5 AND MARINE 4		6		7		8	
	All other	Group R	All other	Group R	All other	Group R	All other	Group R	All other	Group R	All other	Group R	All other	Group R	All other	Group R
Roofs																
Insulation entirely above deck	U-0.048	U-0.048	U-0.048	U-0.048	U-0.048	U-0.048	U-0.039	U-0.039	U-0.039	U-0.039	U-0.032	U-0.032	U-0.028	U-0.028	U-0.028	U-0.028
Metal buildings	U-0.044	U-0.035	U-0.035	U-0.035	U-0.035	U-0.035	U-0.035	U-0.035	U-0.035	U-0.035	U-0.031	U-0.031	U-0.029	U-0.029	U-0.029	U-0.029
Attic and other	U-0.027	U-0.027	U-0.027	U-0.027	U-0.027	U-0.027	U-0.027	U-0.021	U-0.027	U-0.021	U-0.021	U-0.021	U-0.021	U-0.021	U-0.021	U-0.021
Walls, Above Grade																
Mass	U-0.142	U-0.142	U-0.142	U-0.123	U-0.110	U-0.104	U-0.104	U-0.090	U-0.078	U-0.078	U-0.078	U-0.071	U-0.061	U-0.061	U-0.061	U-0.061
Metal building	U-0.079	U-0.079	U-0.079	U-0.079	U-0.079	U-0.052	U-0.052	U-0.052	U-0.052	U-0.052	U-0.052	U-0.052	U-0.052	U-0.039	U-0.052	U-0.039
Metal framed	U-0.077	U-0.077	U-0.077	U-0.064	U-0.064	U-0.064	U-0.064	U-0.064	U-0.064	U-0.064	U-0.064	U-0.057	U-0.064	U-0.052	U-0.045	U-0.045
Wood framed and other	U-0.064	U-0.064	U-0.064	U-0.064	U-0.064	U-0.064	U-0.064	U-0.064	U-0.064	U-0.064	U-0.051	U-0.051	U-0.051	U-0.051	U-0.036	U-0.036
Walls, Below Grade																
Below-grade wall[b]	C-1.140	C-1.140	C-1.140	C-1.140	C-1.140	C-1.140	C-0.119	C-0.119	C-0.119	C-0.119	C-0.119	C-0.119	C-0.092	C-0.092	C-0.092	C-0.092
Floors																
Mass	U-0.322	U-0.322	U-0.107	U-0.087	U-0.076	U-0.076	U-0.076	U-0.074	U-0.074	U-0.064	U-0.064	U-0.057	U-0.055	U-0.051	U-0.055	U-0.051
Joist/framing	U-0.066	U-0.066	U-0.033	U-0.033	U-0.033	U-0.033	U-0.033	U-0.033	U-0.033	U-0.033	U-0.033	U-0.033	U-0.033	U-0.033	U-0.033	U-0.033
Slab-on-Grade Floors																
Unheated slabs	F-0.73	F-0.73	F-0.73	F-0.73	F-0.73	F-0.73	F-0.54	F-0.54	F-0.54	F-0.54	F-0.54	F-0.52	F-0.40	F-0.40	F-0.40	F-0.40
Heated slabs	F-0.70	F-0.70	F-0.70	F-0.70	F-0.70	F-0.70	F-0.65	F-0.58	F-0.58	F-0.58	F-0.58	F-0.58	F-0.55	F-0.55	F-0.55	F-0.55

a. Use of opaque assembly U-factors, C-factors, and F-factors from ANSI/ASHRAE/IESNA 90.1 Appendix A shall be permitted, provided the construction complies with the applicable construction details from ANSI/ASHRAE/IESNA 90.1 Appendix A.

b. Where heated slabs are below grade, below-grade walls shall comply with the F-factor requirements for heated slabs.

TABLE C402.2
OPAQUE THERMAL ENVELOPE REQUIREMENTS[a]

CLIMATE ZONE	1		2		3		4 EXCEPT MARINE		5 AND MARINE 4		6		7		8	
	All Other	Group R	All Other	Group R	All Other	Group R	All Other	Group R	All Other	Group R	All Other	Group R	All Other	Group R	All Other	Group R
Roofs																
Insulation entirely above deck	R-20ci	R-20ci	R-20ci	R-20ci	R-20ci	R-20ci	R-25ci	R-25ci	R-25ci	R-25ci	R-30ci	R-30ci	R-35ci	R-35ci	R-35ci	R-35ci
Metal buildings (with R-5 thermal blocks)[a,b]	R-19 + R-5ci	R-19 + R11 LS	R-19 + R-11 LS	R-19 + R-11 LS	R-19 + R-11 LS	R-19 + R-11 LS	R-19 + R-11 LS	R-19 + R-11 LS	R-19 + R-11 LS	R-19 + R-11 LS	R-25 + R-11 LS	R-25 + R-11 LS	R-30 + R-11 LS	R-30 + R-11 LS	R-30 + R-11 LS	R-30 + R-11 LS
Attic and other	R-38	R-38	R-38	R-38	R-38	R-38	R-38	R-38	R-38	R-49	R-49	R-49	R-49	R-49	R-49	R-49
Walls, Above Grade																
Mass[c]	R-5.7ci[c]	R-5.7ci[c]	R-5.7ci[c]	R-7.6ci	R-7.6ci	R-9.5ci	R-9.5ci	R-11.4ci	R-11.4ci	R-13.3ci	R-13.3ci	R-15.2ci	R-15.2ci	R-15.2ci	R-25ci	R-25ci
Metal building	R-13 + R-6.5ci	R-13 + R-6.5ci	R-13 + R-6.5ci	R-13 + R-13ci	R-13 + R-6.5ci	R-13 + R-13ci	R-13 + R-13ci	R-13 + R-13ci	R-13 + R-13ci	R-13 + R-13ci	R-13 + R-13ci	R-13 + R-13ci	R-13 + R-13ci	R-13 + R-19.5ci	R-13 + R-13ci	R-13 + R-19.5ci
Metal framed	R-13 + R-5ci	R-13 + R-5ci	R-13 + R-5ci	R-13 + R-7.5ci	R-13 + R-7.5ci	R-13 + R-7.5ci	R-13 + R-7.5ci	R-13 + R-7.5ci	R-13 + R-7.5ci	R-13 + R-7.5ci	R-13 + R-7.5ci	R-13 + R-7.5ci	R-13 + R-7.5ci	R-13 + R-15.6ci	R-13 + R-7.5ci	R-13 + R17.5ci
Wood framed and other	R-13 + R-3.8ci or R-20	R-13 + R-3.8ci or R-20	R-13 + R-3.8ci or R-20	R-13 + R-3.8ci or R-20	R-13 + R-3.8ci or R-20	R-13 + R-3.8ci or R-20	R-13 + R-3.8ci or R-20	R-13 + R-3.8ci or R-20	R-13 + R-3.8ci or R-20	R-13 + R-7.5ci or R-20 + R-3.8ci	R-13 + R-7.5ci or R-20 + R-3.8ci	R-13 + R-7.5ci or R-20 + R-3.8ci	R-13 + R-7.5ci or R-20 + R-3.8ci	R-13 + R-7.5ci or R-20 + R-3.8ci	R-13 + R-15.6ci or R-20 + R-10ci	R-13 + R-15.6ci or R-20 + R-10ci
Walls, Below Grade																
Below-grade wall[d]	NR	NR	NR	NR	NR	NR	R-7.5ci	R-7.5ci	R-7.5ci	R-7.5ci	R-7.5ci	R-7.5ci	R-10ci	R-10ci	R-12.5ci	R-12.5ci
Floors																
Mass	NR	NR	R-6.3ci	R-8.3ci	R-10ci	R-10ci	R-10ci	R-10.4ci	R-10ci	R-12.5ci	R-12.5ci	R-12.5ci	R-15ci	R-16.7ci	R-16.7ci	R-16.7ci
Joist/framing	NR	NR	R-30	R-30	R-30	R-30	R-30	R-30	R-30	R-30	R-30	R-30	R-30[e]	R-30[e]	R-30[e]	R-30[e]
Slab-on-Grade Floors																
Unheated slabs	NR	NR	NR	NR	NR	NR	R-10 for 24" below	R-10 for 24" below	R-10 for 24" below	R-10 for 24" below	R-10 for 24" below	R-10 for 24" below	R-15 for 24" below	R-15 for 24" below	R-15 for 24" below	R-20 for 24" below
Heated slabs[d]	R-7.5 for 12" below	R-7.5 for 12" below	R-7.5 for 12" below	R-7.5 for 12" below	R-10 for 24" below	R-10 for 24" below	R-15 for 24" below	R-15 for 24" below	R-15 for 36" below	R-15 for 36" below	R-15 for 36" below	R-20 for 48" below	R-20 for 48" below	R-20 for 48" below	R-20 for 48" below	R-20 for 48" below
Opaque Doors																
Swinging	U-0.61	U-0.61	U-0.61	U-0.61	U-0.61	U-0.61	U-0.61	U-0.61	U-0.37	U-0.37	U-0.37	U-0.37	U-0.37	U-0.37	U-0.37	U-0.37
Roll-up or sliding	R-4.75	R-4.75	R-4.75	R-4.75	R-4.75	R-4.75	R-4.75	R-4.75	R-4.75	R-4.75	R-4.75	R-4.75	R-4.75	R-4.75	R-4.75	R-4.75

For SI: 1 inch = 25.4 mm.   ci = Continuous insulation.   NR = No requirement.

LS = Liner System—A continuous membrane installed below the purlins and uninterrupted by framing members. Uncompressed, unfaced insulation rests on top of the membrane between the purlins.

a. Assembly descriptions can be found in ANSI/ASHRAE/IESNA Appendix A.
b. Where using R-value compliance method, a thermal spacer block shall be provided, otherwise use the U-factor compliance method in Table C402.1.2.
c. R-5.7ci is allowed to be substituted with concrete block walls complying with ASTM C 90, ungrouted or partially grouted at 32 inches or less on center vertically and 48 inches or less on center horizontally, with ungrouted cores filled with materials having a maximum thermal conductivity of 0.44 Btu-in/h-ft^2 °F.
d. Where heated slabs are below grade, below-grade walls shall comply with the exterior insulation requirements for heated slabs.
e. Steel floor joist systems shall be insulated to R-38.

C402.3 Fenestration (Prescriptive). Fenestration shall comply with Table C402.3. Automatic daylighting controls specified by this section shall comply with Section C405.2.2.3.2.

C402.3.1 Maximum area. The vertical fenestration area (not including opaque doors and opaque spandrel panels) shall not exceed 30 percent of the gross above-grade wall area. The skylight area shall not exceed 3 percent of the gross roof area.

C402.3.1.1 Increased vertical fenestration area with daylighting controls. In Climate Zones 1 through 6, a maximum of 40 percent of the gross above-grade wall area shall be permitted to be vertical fenestration, provided:

1. No less than 50 percent of the conditioned floor area is within a daylight zone;

2. Automatic daylighting controls are installed in daylight zones; and

3. Visible transmittance (VT) of vertical fenestration is greater than or equal to 1.1 times solar heat gain coefficient (SHGC).

Exception: Fenestration that is outside the scope of NFRC 200 is not required to comply with Item 3.

C402.3.1.2 Increased skylight area with daylighting controls. The skylight area shall be permitted to be a maximum of 5 percent of the roof area provided automatic daylighting controls are installed in daylight zones under skylights.

C402.3.2 Minimum skylight fenestration area. In an enclosed space greater than 10,000 square feet (929 m²), directly under a roof with ceiling heights greater than 15 feet (4572 mm), and used as an office, lobby, atrium, concourse, corridor, storage, gymnasium/exercise center, convention center, automotive service, manufacturing, non-refrigerated warehouse, retail store, distribution/sorting area, transportation, or workshop, the total daylight zone under skylights shall be not less than half the floor area

and shall provide a minimum skylight area to daylight zone under skylights of either:

1. Not less than 3 percent with a skylight VT of at least 0.40; or

2. Provide a minimum skylight effective aperture of at least 1 percent determined in accordance with Equation C4-1.

$$\text{Skylight Effective Aperture} = \frac{0.85 \times \text{Skylight Area} \times \text{Skylight VT} \times \text{WF}}{\text{Daylight zone under skylight}}$$

(Equation C4-1)

where:

Skylight area = Total fenestration area of skylights.

Skylight VT = Area weighted average visible transmittance of skylights.

WF = Area weighted average well factor, where well factor is 0.9 if light well depth is less than 2 feet (610 mm), or 0.7 if light well depth is 2 feet (610 mm) or greater.

Light well depth = Measure vertically from the underside of the lowest point of the skylight glazing to the ceiling plane under the skylight.

Exception: Skylights above daylight zones of enclosed spaces are not required in:

1. Buildings in climate zones 6 through 8.

2. Spaces where the designed *general lighting* power densities are less than 0.5 W/ft² (5.4 W/m²).

3. Areas where it is documented that existing structures or natural objects block direct beam sunlight on at least half of the roof over the enclosed

TABLE C402.3
BUILDING ENVELOPE REQUIREMENTS: FENESTRATION

CLIMATE ZONE	1	2	3	4 EXCEPT MARINE	5 AND MARINE 4	6	7	8
Vertical fenestration								
U-factor								
Fixed fenestration	0.50	0.50	0.46	0.38	0.38	0.36	0.29	0.29
Operable fenestration	0.65	0.65	0.60	0.45	0.45	0.43	0.37	0.37
Entrance doors	1.10	0.83	0.77	0.77	0.77	0.77	0.77	0.77
SHGC								
SHGC	0.25	0.25	0.25	0.40	0.40	0.40	0.45	0.45
Skylights								
U-factor	0.75	0.65	0.55	0.50	0.50	0.50	0.50	0.50
SHGC	0.35	0.35	0.35	0.40	0.40	0.40	NR	NR

NR = No requirement.

area for more than 1,500 daytime hours per year between 8 am and 4 pm.

4. Spaces where the daylight zone under rooftop monitors is greater than 50 percent of the enclosed space floor area.

C402.3.2.1 Lighting controls in daylight zones under skylights. All lighting in the daylight zone shall be controlled by multilevel lighting controls that comply with Section C405.2.2.3.3.

Exception: Skylights above daylight zones of enclosed spaces are not required in:

1. Buildings in Climate Zones 6 through 8.

2. Spaces where the designed *general lighting* power densities are less than 0.5 W/ft^2 (5.4 W/m^2).

3. Areas where it is documented that existing structures or natural objects block direct beam sunlight on at least half of the roof over the enclosed area for more than 1,500 daytime hours per year between 8 am and 4 pm.

4. Spaces where the daylight zone under rooftop monitors is greater than 50 percent of the enclosed space floor area.

C402.3.2.2 Haze factor. Skylights in office, storage, automotive service, manufacturing, nonrefrigerated warehouse, retail store, and distribution/sorting area spaces shall have a glazing material or diffuser with a measured haze factor greater than 90 percent when tested in accordance with ASTM D 1003.

Exception: Skylights designed to exclude direct sunlight entering the occupied space by the use of fixed or automated baffles, or the geometry of skylight and light well need not comply with Section C402.3.2.2.

C402.3.3 Maximum *U*-factor and SHGC. For vertical fenestration, the maximum *U*-factor and solar heat gain coefficient (SHGC) shall be as specified in Table C402.3, based on the window projection factor. For skylights, the maximum *U*-factor and solar heat gain coefficient (SHGC) shall be as specified in Table C402.3.

The window projection factor shall be determined in accordance with Equation C4-2.

$$PF = A/B \qquad \text{(Equation C4-2)}$$

where:

PF = Projection factor (decimal).

A = Distance measured horizontally from the furthest continuous extremity of any overhang, eave, or permanently attached shading device to the vertical surface of the glazing.

B = Distance measured vertically from the bottom of the glazing to the underside of the overhang, eave, or permanently attached shading device.

Where different windows or glass doors have different *PF* values, they shall each be evaluated separately.

C402.3.3.1 SHGC adjustment. Where the fenestration projection factor for a specific vertical fenestration product is greater than or equal to 0.2, the required maximum SHGC from Table C402.3 shall be adjusted by multiplying the required maximum SHGC by the multiplier specified in Table C402.3.3.1 corresponding with the orientation of the fenestration product and the projection factor.

TABLE C402.3.3.1
SHGC ADJUSTMENT MULTIPLIERS

PROJECTION FACTOR	ORIENTED WITHIN 45 DEGREES OF TRUE NORTH	ALL OTHER ORIENTATION
$0.2 \leq PF < 0.5$	1.1	1.2
$PF \geq 0.5$	1.2	1.6

C402.3.3.2 Increased vertical fenestration SHGC. In Climate Zones 1, 2 and 3, vertical fenestration entirely located not less than 6 feet (1729 mm) above the finished floor shall be permitted a maximum SHGC of 0.40.

C402.3.3.3 Increased skylight SHGC. In Climate Zones 1 through 6, skylights shall be permitted a maximum SHGC of 0.60 where located above daylight zones provided with automated daylighting controls.

C402.3.3.4 Increased skylight *U*-factor. Where skylights are installed above daylight zones provided with automatic daylighting controls, a maximum *U*-factor of 0.9 shall be permitted in Climate Zones 1 through 3; and a maximum *U*-factor of 0.75 shall be permitted in Climate Zones 4 through 8.

C402.3.3.5 Dynamic glazing. For compliance with Section C402.3.3, the SHGC for dynamic glazing shall be determined using the manufacturer's lowest-rated SHGC, and the VT/SHGC ratio shall be determined using the maximum VT and maximum SHGC. Dynamic glazing shall be considered separately from other fenestration, and area-weighted averaging with other fenestration that is not dynamic glazing shall not be permitted.

C402.3.4 Area-weighted *U*-factor. An area-weighted average shall be permitted to satisfy the *U*-factor requirements for each fenestration product category listed in Table C402.3. Individual fenestration products from different fenestration product categories listed in Table C402.3 shall not be combined in calculating area-weighted average *U*-factor.

C402.4 Air leakage (Mandatory). The thermal envelope of buildings shall comply with Sections C402.4.1 through C402.4.8.

C402.4.1 Air barriers. A continuous air barrier shall be provided throughout the building thermal envelope. The air barriers shall be permitted to be located on the inside or outside of the building envelope, located within the assemblies composing the envelope, or any combination thereof. The air barrier shall comply with Sections C402.4.1.1 and C402.4.1.2.

Exception: Air barriers are not required in buildings located in Climate Zones 1, 2 and 3.

C402.4.1.1 Air barrier construction. The *continuous air barrier* shall be constructed to comply with the following:

1. The air barrier shall be continuous for all assemblies that are the thermal envelope of the building and across the joints and assemblies.

2. Air barrier joints and seams shall be sealed, including sealing transitions in places and changes in materials. Air barrier penetrations shall be sealed in accordance with Section C402.4.2. The joints and seals shall be securely installed in or on the joint for its entire length so as not to dislodge, loosen or otherwise impair its ability to resist positive and negative pressure from wind, stack effect and mechanical ventilation.

3. Recessed lighting fixtures shall comply with Section C402.4.8. Where similar objects are installed which penetrate the air barrier, provisions shall be made to maintain the integrity of the air barrier.

Exception: Buildings that comply with Section C402.4.1.2.3 are not required to comply with Items 1 and 3.

C402.4.1.2 Air barrier compliance options. A continuous air barrier for the opaque building envelope shall comply with Section C402.4.1.2.1, C402.4.1.2.2, or C402.4.1.2.3.

C402.4.1.2.1 Materials. Materials with an air permeability no greater than 0.004 cfm/ft^2 (0.02 L/s · m^2) under a pressure differential of 0.3 inches water gauge (w.g.) (75 Pa) when tested in accordance with ASTM E 2178 shall comply with this section. Materials in Items 1 through 15 shall be deemed to comply with this section provided joints are sealed and materials are installed as air barriers in accordance with the manufacturer's instructions.

1. Plywood with a thickness of not less than $^3/_8$ inch (10 mm).

2. Oriented strand board having a thickness of not less than $^3/_8$ inch (10 mm).

3. Extruded polystyrene insulation board having a thickness of not less than $^1/_2$ inch (12 mm).

4. Foil-back polyisocyanurate insulation board having a thickness of not less than $^1/_2$ inch (12 mm).

5. Closed cell spray foam a minimum density of 1.5 pcf (2.4 kg/m^3) having a thickness of not less than $1^1/_2$ inches (36 mm).

6. Open cell spray foam with a density between 0.4 and 1.5 pcf (0.6 and 2.4 kg/m^3) and having a thickness of not less than 4.5 inches (113 mm).

7. Exterior or interior gypsum board having a thickness of not less than $^1/_2$ inch (12 mm).

8. Cement board having a thickness of not less than $^1/_2$ inch (12 mm).

9. Built up roofing membrane.

10. Modified bituminous roof membrane.

11. Fully adhered single-ply roof membrane.

12. A Portland cement/sand parge, or gypsum plaster having a thickness of not less than $^5/_8$ inch (16 mm).

13. Cast-in-place and precast concrete.

14. Fully grouted concrete block masonry.

15. Sheet steel or aluminum.

C402.4.1.2.2 Assemblies. Assemblies of materials and components with an average air leakage not to exceed 0.04 cfm/ft^2 (0.2 L/s · m^2) under a pressure differential of 0.3 inches of water gauge (w.g.)(75 Pa) when tested in accordance with ASTM E 2357, ASTM E 1677 or ASTM E 283 shall comply with this section. Assemblies listed in Items 1 and 2 shall be deemed to comply provided joints are sealed and requirements of Section C402.4.1.1 are met.

1. Concrete masonry walls coated with one application either of block filler and two applications of a paint or sealer coating;

2. A Portland cement/sand parge, stucco or plaster minimum $^1/_2$ inch (12 mm) in thickness.

C402.4.1.2.3 Building test. The completed building shall be tested and the air leakage rate of the *building envelope* shall not exceed 0.40 cfm/ft^2 at a pressure differential of 0.3 inches water gauge (2.0 L/s · m^2 at 75 Pa) in accordance with ASTM E 779 or an equivalent method approved by the code official.

C402.4.2 Air barrier penetrations. Penetrations of the air barrier and paths of air leakage shall be caulked, gasketed or otherwise sealed in a manner compatible with the construction materials and location. Joints and seals shall be sealed in the same manner or taped or covered with a moisture vapor-permeable wrapping material. Sealing materials shall be appropriate to the construction materials being sealed. The joints and seals shall be securely installed in or on the joint for its entire length so as not to dislodge, loosen or otherwise impair its ability to resist positive and negative pressure from wind, stack effect and mechanical ventilation.

C402.4.3 Air leakage of fenestration. The air leakage of fenestration assemblies shall meet the provisions of Table C402.4.3. Testing shall be in accordance with the applicable reference test standard in Table C402.4.3 by an accredited, independent testing laboratory and *labeled* by the manufacturer.

Exceptions:

1. Field-fabricated fenestration assemblies that are sealed in accordance with Section C402.4.1.

2. Fenestration in buildings that comply with Section C402.4.1.2.3 are not required to meet the air leakage requirements in Table C402.4.3.

TABLE C402.4.3
MAXIMUM AIR INFILTRATION RATE
FOR FENESTRATION ASSEMBLIES

FENESTRATION ASSEMBLY	MAXIMUM RATE(CFM/FT²)	TEST PROCEDURE
Windows	0.20 [a]	AAMA/WDMA/CSA101/I.S.2/A440 or NFRC 400
Sliding doors	0.20 [a]	
Swinging doors	0.20 [a]	
Skylights – with condensation weepage openings	0.30	
Skylights – all other	0.20 [a]	
Curtain walls	0.06	NFRC 400 or ASTM E 283 at 1.57 psf (75 Pa)
Storefront glazing	0.06	
Commercial glazed swinging entrance doors	1.00	
Revolving doors	1.00	
Garage doors	0.40	ANSI/DASMA 105, NFRC 400, or ASTM E 283 at 1.57 psf (75 Pa)
Rolling doors	1.00	

For SI: 1 cubic foot per minute = 0.47L/s, 1 square foot = 0.093 m².

a. The maximum rate for windows, sliding and swinging doors, and skylights is permitted to be 0.3 cfm per square foot of fenestration or door area when tested in accordance with AAMA/WDMA/CSA101/I.S.2/A440 at 6.24 psf (300 Pa).

C402.4.4 Doors and access openings to shafts, chutes, stairways, and elevator lobbies. Doors and access openings from conditioned space to shafts, chutes, stairways and elevator lobbies shall either meet the requirements of Section C402.4.3 or shall be gasketed, weatherstripped or sealed.

> **Exception:** Door openings required to comply with Section 716 or 716.5 of the *International Building Code*; or doors and door openings required by the *International Building Code* to comply with UL 1784 shall not be required to comply with Section C402.4.4.

C402.4.5 Air intakes, exhaust openings, stairways and shafts. Stairway enclosures and elevator shaft vents and other outdoor air intakes and exhaust openings integral to the building envelope shall be provided with dampers in accordance with Sections C402.4.5.1 and C402.4.5.2.

> **C402.4.5.1 Stairway and shaft vents.** Stairway and shaft vents shall be provided with Class I motorized dampers with a maximum leakage rate of 4 cfm/ft² (20.3 L/s · m²) at 1.0 inch water gauge (w.g.) (249 Pa) when tested in accordance with AMCA 500D.
>
> Stairway and shaft vent dampers shall be installed with controls so that they are capable of automatically opening upon:
>
> 1. The activation of any fire alarm initiating device of the building's fire alarm system; or
> 2. The interruption of power to the damper.

> **C402.4.5.2 Outdoor air intakes and exhausts.** *Outdoor air* supply and exhaust openings shall be provided with Class IA motorized dampers with a maximum

leakage rate of 4 cfm/ft² (20.3 L/s · m²) at 1.0 inch water gauge (w.g.) (249 Pa) when tested in accordance with AMCA 500D.

> **Exceptions:**
>
> 1. Gravity (nonmotorized) dampers having a maximum leakage rate of 20 cfm/ft² (101.6 L/s · m²) at 1.0 inch water gauge (w.g.) (249 Pa) when tested in accordance with AMCA 500D are permitted to be used as follows:
>
>> 1.1. In buildings for exhaust and relief dampers.
>>
>> 1.2. In buildings less than three stories in height above grade.
>>
>> 1.3. For ventilation air intakes and exhaust and relief dampers in buildings of any height located in Climate Zones 1, 2 and 3.
>>
>> 1.4. Where the design *outdoor air* intake or exhaust capacity does not exceed 300 cfm (141 L/s).
>
>> Gravity (nonmotorized) dampers for ventilation air intakes shall be protected from direct exposure to wind.
>
> 2. Dampers smaller than 24 inches (610 mm) in either dimension shall be permitted to have a leakage of 40 cfm/ft² (203.2 L/s · m²) at 1.0 inch water gauge (w.g.) (249 Pa) when tested in accordance with AMCA 500D.

C402.4.6 Loading dock weatherseals. Cargo doors and loading dock doors shall be equipped with weatherseals to restrict infiltration when vehicles are parked in the doorway.

C402.4.7 Vestibules. All building entrances shall be protected with an enclosed vestibule, with all doors opening into and out of the vestibule equipped with self-closing devices. Vestibules shall be designed so that in passing through the vestibule it is not necessary for the interior and exterior doors to open at the same time. The installation of one or more revolving doors in the building entrance shall not eliminate the requirement that a vestibule be provided on any doors adjacent to revolving doors.

> **Exceptions:**
>
> 1. Buildings in Climate Zones 1 and 2.
> 2. Doors not intended to be used by the public, such as doors to mechanical or electrical equipment rooms, or intended solely for employee use.
> 3. Doors opening directly from a *sleeping unit* or dwelling unit.
> 4. Doors that open directly from a space less than 3,000 square feet (298 m²) in area.
> 5. Revolving doors.

6. Doors used primarily to facilitate vehicular movement or material handling and adjacent personnel doors.

C402.4.8 Recessed lighting. Recessed luminaires installed in the *building thermal envelope* shall be sealed to limit air leakage between conditioned and unconditioned spaces. All recessed luminaires shall be IC-rated and *labeled* as having an air leakage rate of not more than 2.0 cfm (0.944 L/s) when tested in accordance with ASTM E 283 at a 1.57 psf (75 Pa) pressure differential. All recessed luminaires shall be sealed with a gasket or caulk between the housing and interior wall or ceiling covering.

SECTION C403
BUILDING MECHANICAL SYSTEMS

C403.1 General. Mechanical systems and equipment serving the building heating, cooling or ventilating needs shall comply with Section C403.2 (referred to as the mandatory provisions) and either:

1. Section C403.3 (Simple systems); or

2. Section C403.4 (Complex systems).

C403.2 Provisions applicable to all mechanical systems (Mandatory). Mechanical systems and equipment serving the building heating, cooling or ventilating needs shall comply with Sections C403.2.1 through C403.2.11.

C403.2.1 Calculation of heating and cooling loads. Design loads shall be determined in accordance with the procedures described in ANSI/ASHRAE/ACCA Standard 183. The design loads shall account for the building envelope, lighting, ventilation and occupancy loads based on the project design. Heating and cooling loads shall be adjusted to account for load reductions that are achieved where energy recovery systems are utilized in the HVAC system in accordance with the ASHRAE *HVAC Systems and Equipment Handbook*. Alternatively, design loads shall be determined by an *approved* equivalent computation procedure, using the design parameters specified in Chapter 3.

C403.2.2 Equipment and system sizing. The output capacity of heating and cooling equipment and systems shall not exceed the loads calculated in accordance with Section C403.2.1. A single piece of equipment providing both heating and cooling shall satisfy this provision for one function with the capacity for the other function as small as possible, within available equipment options.

Exceptions:

1. Required standby equipment and systems provided with controls and devices that allow such systems or equipment to operate automatically only when the primary equipment is not operating.

2. Multiple units of the same equipment type with combined capacities exceeding the design load and provided with controls that have the capability to sequence the operation of each unit based on load.

C403.2.3 HVAC equipment performance requirements. Equipment shall meet the minimum efficiency requirements of Tables C403.2.3(1), C403.2.3(2), C403.2.3(3), C403.2.3(4), C403.2.3(5), C403.2.3(6), C403.2.3(7) and C403.2.3(8) when tested and rated in accordance with the applicable test procedure. Plate-type liquid-to-liquid heat exchangers shall meet the minimum requirements of Table C403.2.3(9). The efficiency shall be verified through certification under an *approved* certification program or, if no certification program exists, the equipment efficiency ratings shall be supported by data furnished by the manufacturer. Where multiple rating conditions or performance requirements are provided, the equipment shall satisfy all stated requirements. Where components, such as indoor or outdoor coils, from different manufacturers are used, calculations and supporting data shall be furnished by the designer that demonstrates that the combined efficiency of the specified components meets the requirements herein.

C403.2.3.1 Water-cooled centrifugal chilling packages. Equipment not designed for operation at AHRI Standard 550/590 test conditions of 44°F (7°C) leaving chilled-water temperature and 85°F (29°C) entering condenser water temperature with 3 gpm/ton (0.054 I/s · kW) condenser water flow shall have maximum full-load kW/ton and *NPLV* ratings adjusted using Equations C4-3 and C4-4.

Adjusted minimum full-load COP ratings =
(Full-load COP from Table 6.8.1C of AHRI Standard 550/590) × K_{adj}

(Equation C4-3)

Adjusted minimum NPLV rating =
(IPLV from Table 6.8.1C of AHRI Standard 550/590) × K_{adj}

(Equation C4-4)

where:

K_{adj} = A × B

A = $0.0000015318 \times (\text{LIFT})^4 - 0.000202076 \times (\text{LIFT})^3 + 0.0101800 \times (\text{LIFT})^2 - 0.264958 \times \text{LIFT} + 3.930196$

B = $0.0027 \times L_{vg}^{\text{Evap}} (°C) + 0.982$

LIFT = $L_{vg}^{\text{Cond}} - L_{vg}^{\text{Evap}}$

L_{vg}^{Cond} = Full-load condenser leaving water temperature (°C)

L_{vg}^{Evap} = Full-load leaving evaporator temperature (°C)

SI units shall be used in the K_{adj} equation.

The adjusted full-load and *NPLV* values shall only be applicable for centrifugal chillers meeting all of the following full-load design ranges:

1. The leaving evaporator fluid temperature is not less than 36°F (2.2°C).

2. The leaving condenser fluid temperature is not greater than 115°F (46.1°C).

3. LIFT is not less than 20°F (11.1 °C) and not greater than 80°F (44.4°C).

Exception: Centrifugal chillers designed to operate outside of these ranges need not comply with this code.

TABLE C403.2.3(1)
MINIMUM EFFICIENCY REQUIREMENTS:
ELECTRICALLY OPERATED UNITARY AIR CONDITIONERS AND CONDENSING UNITS

EQUIPMENT TYPE	SIZE CATEGORY	HEATING SECTION TYPE	SUBCATEGORY OR RATING CONDITION	MINIMUM EFFICIENCY		TEST PROCEDURE[a]
				Before 6/1/2011	As of 6/1/2011	
Air conditioners, air cooled	< 65,000 Btu/h[b]	All	Split System	13.0 SEER	13.0 SEER	AHRI 210/240
			Single Package	13.0 SEER	13.0 SEER	
Through-the-wall (air cooled)	≤ 30,000 Btu/h[b]	All	Split system	12.0 SEER	12.0 SEER	
			Single Package	12.0 SEER	12.0 SEER	
Small-duct high-velocity (air cooled)	< 65,000 Btu/h[b]	All	Split System	10.0 SEER	10.0 SEER	
Air conditioners, air cooled	≥ 65,000 Btu/h and < 135,000 Btu/h	Electric Resistance (or None)	Split System and Single Package	11.2 EER 11.4 IEER	11.2 EER 11.4 IEER	AHRI 340/360
		All other	Split System and Single Package	11.0 EER 11.2 IEER	11.0 EER 11.2 IEER	
	≥ 135,000 Btu/h and < 240,000 Btu/h	Electric Resistance (or None)	Split System and Single Package	11.0 EER 11.2 IEER	11.0 EER 11.2 IEER	
		All other	Split System and Single Package	10.8 EER 11.0 IEER	10.8 EER 11.0 IEER	
	≥ 240,000 Btu/h and < 760,000 Btu/h	Electric Resistance (or None)	Split System and Single Package	10.0 EER 10.1 IEER	10.0 EER 10.1 IEER	
		All other	Split System and Single Package	9.8 EER 9.9 IEER	9.8 EER 9.9 IEER	
	≥ 760,000 Btu/h	Electric Resistance (or None)	Split System and Single Package	9.7 EER 9.8 IEER	9.7 EER 9.8 IEER	
		All other	Split System and Single Package	9.5 EER 9.6 IEER	9.5 EER 9.6 IEER	
Air conditioners, water cooled	< 65,000 Btu/h[b]	All	Split System and Single Package	12.1 EER 12.3 IEER	12.1 EER 12.3 IEER	AHRI 210/240
	≥ 65,000 Btu/h and < 135,000 Btu/h	Electric Resistance (or None)	Split System and Single Package	11.5 EER 11.7 IEER	12.1 EER 12.3 IEER	AHRI 340/360
		All other	Split System and Single Package	11.3 EER 11.5 IEER	11.9 EER 12.1 IEER	
	≥ 135,000 Btu/h and < 240,000 Btu/h	Electric Resistance (or None)	Split System and Single Package	11.0 EER 11.2 IEER	12.5 EER 12.7 IEER	
		All other	Split System and Single Package	10.8 EER 11.0 IEER	12.3 EER 12.5 IEER	
	≥ 240,000 Btu/h and < 760,000 Btu/h	Electric Resistance (or None)	Split System and Single Package	11.0 EER 11.1 EER	12.4 EER 12.6 EER	
		All other	Split System and Single Package	10.8 EER 10.9 EER	12.2 EER 12.4 EER	
	≥ 760,000 Btu/h	Electric Resistance (or None)	Split System and Single Package	11.0 EER 11.1 EER	12.0 EER 12.4 EER	
		All other	Split System and Single Package	10.8 EER 10.9 EER	12.0 EER 12.2 EER	

(continued)

TABLE C403.2.3(1)—continued
MINIMUM EFFICIENCY REQUIREMENTS:
ELECTRICALLY OPERATED UNITARY AIR CONDITIONERS AND CONDENSING UNITS

EQUIPMENT TYPE	SIZE CATEGORY	HEATING SECTION TYPE	SUB-CATEGORY OR RATING CONDITION	MINIMUM EFFICIENCY		TEST PROCEDURE[a]
				Before 6/1/2011	As of 6/1/2011	
Air conditioners, evaporatively cooled	< 65,000 Btu/h[b]	All	Split System and Single Package	12.1 EER 12.3 IEER	12.1 EER 12.3 IEER	AHRI 210/240
	≥ 65,000 Btu/h and < 135,000 Btu/h	Electric Resistance (or None)	Split System and Single Package	11.5 EER 11.7 IEER	12.1 EER 12.3 IEER	AHRI 340/360
		All other	Split System and Single Package	11.3 EER 11.5 IEER	11.9 EER 12.1 IEER	
	≥ 135,000 Btu/h and < 240,000 Btu/h	Electric Resistance (or None)	Split System and Single Package	11.0 EER 11.2 IEER	12.0 EER 12.2 IEER	
		All other	Split System and Single Package	10.8 EER 11.0 IEER	11.8 EER 12.0 IEER	
	≥ 240,000 Btu/h and < 760,000 Btu/h	Electric Resistance (or None)	Split System and Single Package	11.0 EER 11.1 EER	11.9 EER 12.1 EER	
		All other	Split System and Single Package	10.8 EER 10.9 EER	12.2 EER 11.9 EER	
	≥ 760,000 Btu/h	Electric Resistance (or None)	Split System and Single Package	11.0 EER 11.1 EER	11.7 EER 11.9 EER	
		All other	Split System and Single Package	10.8 EER 10.9 EER	11.5 EER 11.7 EER	
Condensing units, air cooled	≥ 135,000 Btu/h			10.1 EER 11.4 IEER	10.5 EER 14.0 IEER	AHRI 365
Condensing units, water cooled	≥ 135,000 Btu/h			13.1 EER 13.6 IEER	13.5 EER 14.0 IEER	
Condensing units, evaporatively cooled	≥ 135,000 Btu/h			13.1 EER 13.6 IEER	13.5 EER 14.0 IEER	

For SI: 1 British thermal unit per hour = 0.2931 W.

a. Chapter 6 of the referenced standard contains a complete specification of the referenced test procedure, including the reference year version of the test procedure.

b. Single-phase, air-cooled air conditioners less than 65,000 Btu/h are regulated by NAECA. SEER values are those set by NAECA.

TABLE C403.2.3(2)
MINIMUM EFFICIENCY REQUIREMENTS:
ELECTRICALLY OPERATED UNITARY AND APPLIED HEAT PUMPS

EQUIPMENT TYPE	SIZE CATEGORY	HEATING SECTION TYPE	SUBCATEGORY OR RATING CONDITION	MINIMUM EFFICIENCY	TEST PROCEDURE[a]
Air cooled (cooling mode)	< 65,000 Btu/h[b]	All	Split System	13.0 SEER	AHRI 210/240
			Single Packaged	13.0 SEER	
Through-the-wall, air cooled	≤ 30,000 Btu/h[b]	All	Split System	13.0 SEER	
			Single Packaged	13.0 SEER	
Single-duct high-velocity air cooled	< 65,000 Btu/h[b]	All	Split System	10.0 SEER	
Air cooled (cooling mode)	≥ 65,000 Btu/h and < 135,000 Btu/h	Electric Resistance (or None)	Split System and Single Package	11.0 EER 11.2 IEER	AHRI 340/360
		All other	Split System and Single Package	10.8 EER 11.0 IEER	
	≥ 135,000 Btu/h and < 240,000 Btu/h	Electric Resistance (or None)	Split System and Single Package	10.6 EER 10.7 IEER	
		All other	Split System and Single Package	10.4 EER 10.5 IEER	
	≥ 240,000 Btu/h	Electric Resistance (or None)	Split System and Single Package	9.5 EER 9.6 IEER	
		All other	Split System and Single Package	9.3 EER 9.4 IEER	
Water source (cooling mode)	< 17,000 Btu/h	All	86°F entering water	11.2 EER	ISO 13256-1
	≥ 17,000 Btu/h and < 65,000 Btu/h	All	86°F entering water	12.0 EER	
	≥ 65,000 Btu/h and < 135,000 Btu/h	All	86°F entering water	12.0 EER	
Ground water source (cooling mode)	< 135,000 Btu/h	All	59°F entering water	16.2 EER	
		All	77°F entering water	13.4 EER	
Water-source water to water (cooling mode)	< 135,000 Btu/h	All	86°F entering water	10.6 EER	ISO 13256-2
			59°F entering water	16.3 EER	
Ground water source Brine to water (cooling mode)	< 135,000 Btu/h	All	77°F entering fluid	12.1 EER	
Air cooled (heating mode)	< 65,000 Btu/h[b]	—	Split System	7.7 HSPF	AHRI 210/240
		—	Single Package	7.7 HSPF	
Through-the-wall, (air cooled, heating mode)	≤ 30,000 Btu/h[b] (cooling capacity)	—	Split System	7.4 HSPF	
		—	Single Package	7.4 HSPF	
Small-duct high velocity (air cooled, heating mode)	< 65,000 Btu/h[b]	—	Split System	6.8 HSPF	

(continued)

TABLE C403.2.3(2)—continued
MINIMUM EFFICIENCY REQUIREMENTS:
ELECTRICALLY OPERATED UNITARY AND APPLIED HEAT PUMPS

EQUIPMENT TYPE	SIZE CATEGORY	HEATING SECTION TYPE	SUB-CATEGORY OR RATING CONDITION	MINIMUM EFFICIENCY	TEST PROCEDURE[a]
Air cooled (heating mode)	≥ 65,000 Btu/h and < 135,000 Btu/h (cooling capacity)	—	47°F db/43°F wb Outdoor Air	3.3 COP	AHRI 340/360
			17°F db/15°F wb Outdoor Air	2.25 COP	
	≥ 135,000 Btu/h (cooling capacity)	—	47°F db/43°F wb Outdoor Air	3.2 COP	
			17°F db/15°F wb Outdoor Air	2.05 COP	
Water source (heating mode)	< 135,000 Btu/h (cooling capacity)	—	68°F entering water	4.2 COP	ISO 13256-1
Ground water source (heating mode)	< 135,000 Btu/h (cooling capacity)	—	50°F entering water	3.6 COP	
Ground source (heating mode)	< 135,000 Btu/h (cooling capacity)	—	32°F entering fluid	3.1 COP	
Water-source water to water (heating mode)	< 135,000 Btu/h (cooling capacity)	—	68°F entering water	3.7 COP	ISO 13256-2
		—	50°F entering water	3.1 COP	
Ground source brine to water (heating mode)	< 135,000 Btu/h (cooling capacity)	—	32°F entering fluid	2.5 COP	

For SI: 1 British thermal unit per hour = 0.2931 W, °C = [(°F) - 32]/1.8.

a. Chapter 6 of the referenced standard contains a complete specification of the referenced test procedure, including the reference year version of the test procedure.

b. Single-phase, air-cooled air conditioners less than 65,000 Btu/h are regulated by NAECA. SEER values are those set by NAECA.

TABLE C403.2.3(3)
MINIMUM EFFICIENCY REQUIREMENTS:
ELECTRICALLY OPERATED PACKAGED TERMINAL AIR CONDITIONERS,
PACKAGED TERMINAL HEAT PUMPS, SINGLE-PACKAGE VERTICAL AIR CONDITIONERS,
SINGLE VERTICAL HEAT PUMPS, ROOM AIR CONDITIONERS AND ROOM AIR-CONDITIONER HEAT PUMPS

EQUIPMENT TYPE	SIZE CATEGORY (INPUT)	SUBCATEGORY OR RATING CONDITION	MINIMUM EFFICIENCY		TEST PROCEDURE[a]
			Before 10/08/2012	As of 10/08/2012	
PTAC (cooling mode) new construction	All Capacities	95°F db outdoor air	12.5 - (0.213 × Cap/1000) EER	13.8 - (0.300 × Cap/1000) EER	AHRI 310/380
PTAC (cooling mode) replacements[b]	All Capacities	95°F db outdoor air	10.9 - (0.213 × Cap/1000) EER	10.9 - (0.213 × Cap/1000) EER	
PTHP (cooling mode) new construction	All Capacities	95°F db outdoor air	12.3 - (0.213 × Cap/1000) EER	14.0 - (0.300 × Cap/1000) EER	
PTHP (cooling mode) replacements[b]	All Capacities	95°F db outdoor air	10.8 - (0.213 × Cap/1000) EER	10.8 - (0.213 × Cap/1000) EER	
PTHP (heating mode) new construction	All Capacities	—	3.2 - (0.026 × Cap/1000) COP	3.2 - (0.026 × Cap/1000) COP	
PTHP (heating mode) replacements[b]	All Capacities	—	2.9 - (0.026 × Cap/1000) COP	2.9 - (0.026 × Cap/1000) COP	
SPVAC (cooling mode)	< 65,000 Btu/h	95°F db/ 75°F wb outdoor air	9.0 EER	9.0 EER	AHRI 390
	≥ 65,000 Btu/h and < 135,000 Btu/h	95°F db/ 75°F wb outdoor air	8.9 EER	8.9 EER	
	≥ 135,000 Btu/h and < 240,000 Btu/h	95°F db/ 75°F wb outdoor air	8.6 EER	8.6 EER	
SPVHP (cooling mode)	< 65,000 Btu/h	95°F db/ 75°F wb outdoor air	9.0 EER	9.0 EER	
	≥ 65,000 Btu/h and < 135,000 Btu/h	95°F db/ 75°F wb outdoor air	8.9 EER	8.9 EER	
	≥ 135,000 Btu/h and < 240,000 Btu/h	95°F db/ 75°F wb outdoor air	8.6 EER	8.6 EER	
SPVHP (heating mode)	<65,000 Btu/h	47°F db/ 43°F wb outdoor air	3.0 COP	3.0 COP	AHRI 390
	≥ 65,000 Btu/h and < 135,000 Btu/h	47°F db/ 43°F wb outdoor air	3.0 COP	3.0 COP	
	≥ 135,000 Btu/h and < 240,000 Btu/h	47°F db/ 75°F wb outdoor air	2.9 COP	2.9 COP	

(continued)

TABLE C403.2.3(3)—continued
MINIMUM EFFICIENCY REQUIREMENTS:
ELECTRICALLY OPERATED PACKAGED TERMINAL AIR CONDITIONERS,
PACKAGED TERMINAL HEAT PUMPS, SINGLE-PACKAGE VERTICAL AIR CONDITIONERS,
SINGLE VERTICAL HEAT PUMPS, ROOM AIR CONDITIONERS AND ROOM AIR-CONDITIONER HEAT PUMPS

EQUIPMENT TYPE	SIZE CATEGORY (INPUT)	SUBCATEGORY OR RATING CONDITION	MINIMUM EFFICIENCY		TEST PROCEDURE[a]
			Before 10/08/2012	As of 10/08/2012	
Room air conditioners, with louvered slides	< 6,000 Btu/h	—	9.7 SEER	9.7 SEER	ANSI/AHA-MRAC-1
	≥ 6,000 Btu/h and < 8,000 Btu/h	—	9.7 EER	9.7 EER	
	≥ 8,000 Btu/h and < 14,000 Btu/h	—	9.8 EER	9.8 EER	
	≥ 14,000 Btu/h and < 20,000 Btu/h	—	9.7 SEER	9.7 SEER	
	≥ 20,000 Btu/h	—	8.5 EER	8.5 EER	
Room air conditioners, without louvered slides	< 8,000 Btu/h	—	9.0 EER	9.0 EER	
	≥ 8,000 Btu/h and < 20,000 Btu/h	—	8.5 EER	8.5 EER	
	≥ 20,000 Btu/h	—	8.5 EER	8.5 EER	
Room air-conditioner heat pumps with louvered sides	< 20,000 Btu/h	—	9.0 EER	9.0 EER	
	≥ 20,000 Btu/h	—	8.5 EER	8.5 EER	
Room air-conditioner heat pumps without louvered sides	< 14,000 Btu/h	—	8.5 EER	8.5 EER	
	≥ 14,000 Btu/h	—	8.0 EER	8.0 EER	
Room air conditioner casement only	All capacities	—	8.7 EER	8.7 EER	
Room air conditioner casement-slider	All capacities	—	9.5 EER	9.5 EER	

For SI: 1 British thermal unit per hour = 0.2931 W, °C = [(°F) - 32]/1.8.

"Cap" = The rated cooling capacity of the project in Btu/h. If the unit's capacity is less than 7000 Btu/h, use 7000 Btu/h in the calculation. If the unit's capacity is greater than 15,000 Btu/h, use 15,000 Btu/h in the calculations.

a. Chapter 6 of the referenced standard contains a complete specification of the referenced test procedure, including the referenced year version of the test procedure.

b. Replacement unit shall be factory labeled as follows: "MANUFACTURED FOR REPLACEMENT APPLICATIONS ONLY: NOT TO BE INSTALLED IN NEW CONSTRUCTION PROJECTS." Replacement efficiencies apply only to units with existing sleeves less than 16 inches (406 mm) in height and less than 42 inches (1067 mm) in width.

TABLE 403.2.3(4)
WARM AIR FURNACES AND COMBINATION WARM AIR FURNACES/AIR-CONDITIONING UNITS,
WARM AIR DUCT FURNACES AND UNIT HEATERS, MINIMUM EFFICIENCY REQUIREMENTS

EQUIPMENT TYPE	SIZE CATEGORY (INPUT)	SUBCATEGORY OR RATING CONDITION	MINIMUM EFFICIENCY[d, e]	TEST PROCEDURE[a]
Warm air furnaces, gas fired	< 225,000 Btu/h	—	78% AFUE or 80%E_t^c	DOE 10 CFR Part 430 or ANSI Z21.47
	≥ 225,000 Btu/h	Maximum capacity[c]	80%E_t^f	ANSI Z21.47
Warm air furnaces, oil fired	< 225,000 Btu/h	—	78% AFUE or 80%E_t^c	DOE 10 CFR Part 430 or UL 727
	≥ 225,000 Btu/h	Maximum capacity[b]	81%E_t^g	UL 727
Warm air duct furnaces, gas fired	All capacities	Maximum capacity[b]	80%E_c	ANSI Z83.8
Warm air unit heaters, gas fired	All capacities	Maximum capacity[b]	80%E_c	ANSI Z83.8
Warm air unit heaters, oil fired	All capacities	Maximum capacity[b]	80%E_c	UL 731

For SI: 1 British thermal unit per hour = 0.2931 W.

a. Chapter 6 of the referenced standard contains a complete specification of the referenced test procedure, including the referenced year version of the test procedure.

b. Minimum and maximum ratings as provided for and allowed by the unit's controls.

c. Combination units not covered by the National Appliance Energy Conservation Act of 1987 (NAECA) (3-phase power or cooling capacity greater than or equal to 65,000 Btu/h [19 kW]) shall comply with either rating.

d. E_t = Thermal efficiency. See test procedure for detailed discussion.

e. E_c = Combustion efficiency (100% less flue losses). See test procedure for detailed discussion.

f. E_c = Combustion efficiency. Units must also include an IID, have jackets not exceeding 0.75 percent of the input rating, and have either power venting or a flue damper. A vent damper is an acceptable alternative to a flue damper for those furnaces where combustion air is drawn from the conditioned space.

g. E_t = Thermal efficiency. Units must also include an IID, have jacket losses not exceeding 0.75 percent of the input rating, and have either power venting or a flue damper. A vent damper is an acceptable alternative to a flue damper for those furnaces where combustion air is drawn from the conditioned space.

TABLE C403.2.3(5)
MINIMUM EFFICIENCY REQUIREMENTS: GAS- AND OIL-FIRED BOILERS

EQUIPMENT TYPE[a]	SUBCATEGORY OR RATING CONDITION	SIZE CATEGORY (INPUT)	MINIMUM EFFICIENCY	TEST PROCEDURE
Boilers, hot water	Gas-fired	< 300,000 Btu/h	80% AFUE	10 CFR Part 430
		≥ 300,000 Btu/h and ≤ 2,500,000 Btu/h[b]	80% E_t	10CFR Part 431
		> 2,500,00 Btu/h[a]	82% E_c	
	Oil-fired[c]	< 300,000 Btu/h	80% AFUE	10 CFR Part 430
		≥ 300,000 Btu/h and ≤ 2,500,000 Btu/h[b]	82% E_t	10 CFR Part 431
		> 2,500,000 Btu/h[a]	84% E_c	
Boilers, steam	Gas-fired	< 300,000 Btu/h	75% AFUE	10 CFR Part 430
	Gas-fired- all, except natural draft	≥ 300,000 Btu/h and ≤ 2,500,000 Btu/h[b]	79% E_t	10CFR Part 431
		> 2,500,000 Btu/h[a]	79% E_t	
	Gas-fired-natural draft	≥ 300,000 Btu/h and ≤ 2,500,000 Btu/h[b]	77% E_t	
		> 2,500,000 Btu/h[a]	77% E_t	
	Oil-fired[c]	< 300,000 Btu/h	80% AFUE	10 CFR Part 430
		≥ 300,000 Btu/h and ≤ 2,500,000 Btu/h[b]	81% E_t	10CFR Part 431
		> 2,500,000 Btu/h[a]	81% E_t	

For SI: 1 British thermal unit per hour = 0.2931 W.

E_c = Combustion efficiency (100 percent less flue losses). E_t = Thermal efficiency. See referenced standard document for detailed information.

a. These requirements apply to boilers with rated input of 8,000,000 Btu/h or less that are not packaged boilers and to all packaged boilers. Minimum efficiency requirements for boilers cover all capacities of packaged boilers.

b. Maximum capacity – minimum and maximum ratings as provided for and allowed by the unit's controls.

c. Includes oil-fired (residual).

TABLE C403.2.3(6)
MINIMUM EFFICIENCY REQUIREMENTS:
CONDENSING UNITS, ELECTRICALLY OPERATED

EQUIPMENT TYPE	SIZE CATEGORY	MINIMUM EFFICIENCY[b]	TEST PROCEDURE[a]
Condensing units, air cooled	≥ 135,000 Btu/h	10.1 EER 11.2 IPLV	AHRI 365
Condensing units, water or evaporatively cooled	≥ 135,000 Btu/h	13.1 EER 13.1 IPLV	

For SI: 1 British thermal unit per hour = 0.2931 W.

a. Chapter 6 of the referenced standard contains a complete specification of the referenced test procedure, including the referenced year version of the test procedure.

b. IPLVs are only applicable to equipment with capacity modulation.

TABLE C403.2.3(7)
MINIMUM EFFICIENCY REQUIREMENTS:
WATER CHILLING PACKAGES[a]

| EQUIPMENT TYPE | SIZE CATEGORY | UNITS | BEFORE 1/1/2010 | | AS OF 1/1/2010[b] | | | | TEST PROCEDURE[c] |
| | | | | | PATH A | | PATH B | | |
			FULL LOAD	IPLV	FULL LOAD	IPLV	FULL LOAD	IPLV	
Air-cooled chillers	< 150 tons	EER	≥ 9.562	≥ 10.416	≥ 9.562	≥ 12.500	NA	NA	AHRI 550/590
	≥ 150 tons	EER			≥ 9.562	≥ 12.750	NA	NA	
Air cooled without condenser, electrical operated	All capacities	EER	≥ 10.586	≥ 11.782	Air-cooled chillers without condensers shall be rated with matching condensers and comply with the air-cooled chiller efficiency requirements				
Water cooled, electrically operated, reciprocating	All capacities	kW/ton	≤ 0.837	≤ 0.696	Reciprocating units shall comply with water cooled positive displacement efficiency requirements				
Water cooled, electrically operated, positive displacement	< 75 tons	kW/ton	≤ 0.790	≤ 0.676	≤ 0.780	≤ 0.630	≤ 0.800	≤ 0.600	
	≥ 75 tons and < 150 tons	kW/ton			≤ 0.775	≤ 0.615	≤ 0.790	≤ 0.586	
	≥ 150 tons and < 300 tons	kW/ton	≤ 0.717	≤ 0.627	≤ 0.680	≤ 0.580	≤ 0.718	≤ 0.540	
	≥ 300 tons	kW/ton	≤ 0.639	≤ 0.571	≤ 0.620	≤ 0.540	≤ 0.639	≤ 0.490	
Water cooled, electrically operated, centrifugal	< 150 tons	kW/ton	≤ 0.703	≤ 0.669	≤ 0.634	≤ 0.596	≤ 0.639	≤ 0.450	
	≥ 150 tons and < 300 tons	kW/ton	≤ 0.634	≤ 0.596					
	≥ 300 tons and < 600 tons	kW/ton	≤ 0.576	≤ 0.549	≤ 0.576	≤ 0.549	≤ 0.600	≤ 0.400	
	≥ 600 tons	kW/ton	≤ 0.576	≤ 0.549	≤ 0.570	≤ 0.539	≤ 0.590	≤ 0.400	
Air cooled, absorption single effect	All capacities	COP	≥ 0.600	NR	≥ 0.600	NR	NA	NA	AHRI 560
Water cooled, absorption single effect	All capacities	COP	≥ 0.700	NR	≥ 0.700	NR	NA	NA	
Absorption double effect, indirect fired	All capacities	COP	≥ 1.000	≥ 1.050	≥ 1.000	≥ 1.050	NA	NA	
Absorption double effect, direct fired	All capacities	COP	≥ 1.000	≥ 1.000	≥ 1.000	≥ 1.000	NA	NA	

For SI: 1 ton = 3517 W, 1 British thermal unit per hour = 0.2931 W, °C = [(°F) - 32]/1.8.

NA = Not applicable, not to be used for compliance; NR = No requirement.

a. The centrifugal chiller equipment requirements, after adjustment in accordance with Section C403.2.3.1 or Section C403.2.3.2, do not apply to chillers used in low-temperature applications where the design leaving fluid temperature is less than 36°F. The requirements do not apply to positive displacement chillers with leaving fluid temperatures less than or equal to 32°F. The requirements do not apply to absorption chillers with design leaving fluid temperatures less than 40°F.

b. Compliance with this standard can be obtained by meeting the minimum requirements of Path A or B. However, both the full load and IPLV shall be met to fulfill the requirements of Path A or B.

c. Chapter 6 of the referenced standard contains a complete specification of the referenced test procedure, including the referenced year version of the test procedure.

TABLE C403.2.3(8)
MINIMUM EFFICIENCY REQUIREMENTS:
HEAT REJECTION EQUIPMENT

EQUIPMENT TYPE[a]	TOTAL SYSTEM HEAT REJECTION CAPACITY AT RATED CONDITIONS	SUBCATEGORY OR RATING CONDITION	PERFORMANCE REQUIRED[b, c, d]	TEST PROCEDURE[e, f]
Propeller or axial fan open circuit cooling towers	All	95°F Entering Water 85°F Leaving Water 75°F Entering wb	≥ 38.2 gpm/hp	CTI ATC-105 and CTI STD-201
Centrifugal fan open circuit cooling towers	All	95°F Entering Water 85°F Leaving Water 75°F Entering wb	≥ 20.0 gpm/hp	CTI ATC-105 and CTI STD-201
Propeller or axial fan closed circuit cooling towers	All	102°F Entering Water 90°F Leaving Water 75°F Entering wb	≥ 14.0 gpm/hp	CTI ATC-105S and CTI STD-201
Centrifugal closed circuit cooling towers	All	102°F Entering Water 90°F Leaving Water 75°F Entering wb	≥ 7.0 gpm/hp	CTI ATC-105S and CTI STD-201
Air-cooled condensers	All	125°F Condensing Temperature R-22 Test Fluid 190°F Entering Gas Temperature 15°F Subcooling 95°F Entering db	≥ 176,000 Btu/h·hp	ARI 460

For SI: °C = [(°F)-32]/1.8, L/s · kW = (gpm/hp)/(11.83), COP = (Btu/h · hp)/(2550.7).

db = dry bulb temperature, °F, wb = wet bulb temperature, °F.

a. The efficiencies and test procedures for both open and closed circuit cooling towers are not applicable to hybrid cooling towers that contain a combination of wet and dry heat exchange sections.

b. For purposes of this table, open circuit cooling tower performance is defined as the water flow rating of the tower at the thermal rating condition listed in Table 403.2.3(8) divided by the fan nameplate rated motor power.

c. For purposes of this table, closed circuit cooling tower performance is defined as the water flow rating of the tower at the thermal rating condition listed in Table 403.2.3(8) divided by the sum of the fan nameplate rated motor power and the spray pump nameplate rated motor power.

d. For purposes of this table, air-cooled condenser performance is defined as the heat rejected from the refrigerant divided by the fan nameplate rated motor power.

e. Chapter 6 of the referenced standard contains a complete specification of the referenced test procedure, including the referenced year version of the test procedure.

f. If a certification program exists for a covered product, and it includes provisions for verification and challenge of equipment efficiency ratings, then the product shall be listed in the certification program, or, if a certification program exists for a covered product, and it includes provisions for verification and challenge of equipment efficiency ratings, but the product is not listed in the existing certification program, the ratings shall be verified by an independent laboratory test report.

TABLE C403.2.3(9)
HEAT TRANSFER EQUIPMENT

EQUIPMENT TYPE	SUBCATEGORY	MINIMUM EFFICIENCY	TEST PROCEDURE[a]
Liquid-to-liquid heat exchangers	Plate type	NR	AHRI 400

NR = No Requirement

a. Chapter 6 of the referenced standard contains a complete specification of the referenced test procedure, including the referenced year version of the test procedure.

C403.2.3.2 Positive displacement (air- and water-cooled) chilling packages. Equipment with a leaving fluid temperature higher than 32°F (0°C), shall meet the requirements of Table C403.2.3(7) when tested or certified with water at standard rating conditions, in accordance with the referenced test procedure.

C403.2.4 HVAC system controls. Each heating and cooling system shall be provided with thermostatic controls as specified in Section C403.2.4.1, C403.2.4.2, C403.2.4.3, C403.2.4.4, C403.4.1, C403.4.2, C403.4.3 or C403.4.4.

C403.2.4.1 Thermostatic controls. The supply of heating and cooling energy to each *zone* shall be controlled by individual thermostatic controls capable of responding to temperature within the *zone*. Where humidification or dehumidification or both is provided, at least one humidity control device shall be provided for each humidity control system.

Exception: Independent perimeter systems that are designed to offset only building envelope heat losses or gains or both serving one or more perimeter *zones* also served by an interior system provided:

1. The perimeter system includes at least one thermostatic control *zone* for each building exposure having exterior walls facing only one orientation (within +/-45 degrees) (0.8 rad) for more than 50 contiguous feet (15 240 mm); and

2. The perimeter system heating and cooling supply is controlled by a thermostats located within the *zones* served by the system.

C403.2.4.1.1 Heat pump supplementary heat. Heat pumps having supplementary electric resistance heat shall have controls that, except during defrost, prevent supplementary heat operation where the heat pump can meet the heating load.

C403.2.4.2 Set point overlap restriction. Where used to control both heating and cooling, *zone* thermostatic controls shall provide a temperature range or deadband of at least 5°F (2.8°C) within which the supply of heating and cooling energy to the *zone* is capable of being shut off or reduced to a minimum.

Exception: Thermostats requiring manual changeover between heating and cooling modes.

C403.2.4.3 Off-hour controls. Each *zone* shall be provided with thermostatic setback controls that are controlled by either an automatic time clock or programmable control system.

Exceptions:

1. *Zones* that will be operated continuously.

2. *Zones* with a full HVAC load demand not exceeding 6,800 Btu/h (2 kW) and having a readily accessible manual shutoff switch.

C403.2.4.3.1 Thermostatic setback capabilities. Thermostatic setback controls shall have the capability to set back or temporarily operate the system to maintain *zone* temperatures down to 55°F (13°C) or up to 85°F (29°C).

C403.2.4.3.2 Automatic setback and shutdown capabilities. Automatic time clock or programmable controls shall be capable of starting and stopping the system for seven different daily schedules per week and retaining their programming and time setting during a loss of power for at least 10 hours. Additionally, the controls shall have a manual override that allows temporary operation of the system for up to 2 hours; a manually operated timer capable of being adjusted to operate the system for up to 2 hours; or an occupancy sensor.

C403.2.4.3.3 Automatic start capabilities. Automatic start controls shall be provided for each HVAC system. The controls shall be capable of automatically adjusting the daily start time of the HVAC system in order to bring each space to the desired occupied temperature immediately prior to scheduled occupancy.

C403.2.4.4 Shutoff damper controls. Both outdoor air supply and exhaust ducts shall be equipped with motorized dampers that will automatically shut when the systems or spaces served are not in use.

Exceptions:

1. Gravity dampers shall be permitted in buildings less than three stories in height.

2. Gravity dampers shall be permitted for buildings of any height located in Climate Zones 1, 2 and 3.

3. Gravity dampers shall be permitted for outside air intake or exhaust airflows of 300 cfm (0.14 m³/s) or less.

C403.2.4.5 Snow melt system controls. Snow- and ice-melting systems, supplied through energy service to the building, shall include automatic controls capable of shutting off the system when the pavement temperature is above 50°F (10°C) and no precipitation is falling and an automatic or manual control that will allow shutoff when the outdoor temperature is above 40°F (4°C) so that the potential for snow or ice accumulation is negligible.

C403.2.5 Ventilation. Ventilation, either natural or mechanical, shall be provided in accordance with Chapter 4 of the *International Mechanical Code*. Where mechanical ventilation is provided, the system shall provide the capability to reduce the outdoor air supply to the minimum required by Chapter 4 of the *International Mechanical Code*.

C403.2.5.1 Demand controlled ventilation. Demand control ventilation (DCV) shall be provided for spaces larger than 500 square feet (50 m²) and with an average occupant load of 25 people per 1000 square feet (93 m²) of floor area (as established in Table 403.3 of the *International Mechanical Code*) and served by systems with one or more of the following:

1. An air-side economizer;

2. Automatic modulating control of the outdoor air damper; or

3. A design outdoor airflow greater than 3,000 cfm (1400 L/s).

Exception: Demand control ventilation is not required for systems and spaces as follows:

1. Systems with energy recovery complying with Section C403.2.6.

2. Multiple-*zone* systems without direct digital control of individual *zones* communicating with a central control panel.

3. System with a design outdoor airflow less than 1,200 cfm (600 L/s).

4. Spaces where the supply airflow rate minus any makeup or outgoing transfer air requirement is less than 1,200 cfm (600 L/s).

5. Ventilation provided for process loads only.

C403.2.6 Energy recovery ventilation systems. Where the supply airflow rate of a fan system exceeds the values specified in Table C403.2.6, the system shall include an energy recovery system. The energy recovery system shall have the capability to provide a change in the enthalpy of the outdoor air supply of not less than 50 percent of the difference between the outdoor air and return air enthalpies, at design conditions. Where an air economizer is required, the energy recovery system shall include a bypass or controls which permit operation of the economizer as required by Section C403.4

Exception: An energy recovery ventilation system shall not be required in any of the following conditions:

1. Where energy recovery systems are prohibited by the *International Mechanical Code.*

2. Laboratory fume hood systems that include at least one of the following features:

    2.1. Variable-air-volume hood exhaust and room supply systems capable of reducing exhaust and makeup air volume to 50 percent or less of design values.

    2.2. Direct makeup (auxiliary) air supply equal to at least 75 percent of the exhaust rate, heated no warmer than 2°F (1.1°C) above room setpoint, cooled to no cooler than 3°F (1.7°C) below room setpoint, no humidification added, and no simultaneous heating and cooling used for dehumidification control.

3. Systems serving spaces that are heated to less than 60°F (15.5°C) and are not cooled.

4. Where more than 60 percent of the outdoor heating energy is provided from site-recovered or site solar energy.

5. Heating energy recovery in Climate Zones 1 and 2.

6. Cooling energy recovery in Climate Zones 3C, 4C, 5B, 5C, 6B, 7 and 8.

7. Systems requiring dehumidification that employ energy recovery in series with the cooling coil.

8. Where the largest source of air exhausted at a single location at the building exterior is less than 75 percent of the design *outdoor air* flow rate.

9. Systems expected to operate less than 20 hours per week at the *outdoor air* percentage covered by Table C403.2.6

C403.2.7 Duct and plenum insulation and sealing. All supply and return air ducts and plenums shall be insulated with a minimum of R-6 insulation where located in unconditioned spaces and a minimum of R-8 insulation where located outside the building. Where located within a building envelope assembly, the duct or plenum shall be separated from the building exterior or unconditioned or exempt spaces by a minimum of R-8 insulation.

Exceptions:

1. Where located within equipment.

2. Where the design temperature difference between the interior and exterior of the duct or plenum does not exceed 15°F (8°C).

All ducts, air handlers and filter boxes shall be sealed. Joints and seams shall comply with Section 603.9 of the *International Mechanical Code.*

C403.2.7.1 Duct construction. Ductwork shall be constructed and erected in accordance with the *International Mechanical Code.*

C403.2.7.1.1 Low-pressure duct systems. All longitudinal and transverse joints, seams and connections of supply and return ducts operating at a static pressure less than or equal to 2 inches water gauge (w.g.) (500 Pa) shall be securely fastened and sealed with welds, gaskets, mastics (adhesives), mastic-plus-embedded-fabric systems or tapes installed in accordance with the manufacturer's installation instructions. Pressure classifications specific to the duct system shall be clearly indicated on the construction documents in accordance with the *International Mechanical Code.*

Exception: Continuously welded and locking-type longitudinal joints and seams on ducts operating at static pressures less than 2 inches water gauge (w.g.) (500 Pa) pressure classification.

C403.2.7.1.2 Medium-pressure duct systems. All ducts and plenums designed to operate at a static pressure greater than 2 inches water gauge (w.g.) (500 Pa) but less than 3 inches w.g. (750 Pa) shall be insulated and sealed in accordance with Section C403.2.7. Pressure classifications specific to the duct system shall be

TABLE C403.2.6
ENERGY RECOVERY REQUIREMENT

CLIMATE ZONE	PERCENT (%) OUTDOOR AIR AT FULL DESIGN AIRFLOW RATE					
	≥ 30% and < 40%	≥ 40% and < 50%	≥ 50% and < 60%	≥ 60% and < 70%	≥ 70% and < 80%	≥ 80%
	DESIGN SUPPLY FAN AIRFLOW RATE (cfm)					
3B, 3C, 4B, 4C, 5B	NR	NR	NR	NR	≥ 5000	≥ 5000
1B, 2B, 5C	NR	NR	≥ 26000	≥ 12000	≥ 5000	≥ 4000
6B	≥ 11000	≥ 5500	≥ 4500	≥ 3500	≥ 2500	≥ 1500
1A, 2A, 3A, 4A, 5A, 6A	≥ 5500	≥ 4500	≥ 3500	≥ 2000	≥ 1000	> 0
7, 8	≥ 2500	≥ 1000	> 0	> 0	> 0	> 0

NR = not required

clearly indicated on the construction documents in accordance with the *International Mechanical Code*.

C403.2.7.1.3 High-pressure duct systems. Ducts designed to operate at static pressures in excess of 3 inches water gauge (w.g.) (750 Pa) shall be insulated and sealed in accordance with Section C403.2.7. In addition, ducts and plenums shall be leak-tested in accordance with the SMACNA *HVAC Air Duct Leakage Test Manual* with the rate of air leakage (*CL*) less than or equal to 6.0 as determined in accordance with Equation C4-5.

$$CL = F/P^{0.65}$$ **(Equation C4-5)**

where:

F = The measured leakage rate in cfm per 100 square feet of duct surface.

P = The static pressure of the test.

Documentation shall be furnished by the designer demonstrating that representative sections totaling at least 25 percent of the duct area have been tested and that all tested sections meet the requirements of this section.

C403.2.8 Piping insulation. All piping serving as part of a heating or cooling system shall be thermally insulated in accordance with Table C403.2.8.

Exceptions:

1. Factory-installed piping within HVAC equipment tested and rated in accordance with a test procedure referenced by this code.

2. Factory-installed piping within room fan-coils and unit ventilators tested and rated according to AHRI 440 (except that the sampling and variation provisions of Section 6.5 shall not apply) and 840, respectively.

3. Piping that conveys fluids that have a design operating temperature range between 60°F (15°C) and 105°F (41°C).

4. Piping that conveys fluids that have not been heated or cooled through the use of fossil fuels or electric power.

5. Strainers, control valves, and balancing valves associated with piping 1 inch (25 mm) or less in diameter.

6. Direct buried piping that conveys fluids at or below 60°F (15°C)

C403.2.8.1 Protection of piping insulation. Piping insulation exposed to weather shall be protected from damage, including that due to sunlight, moisture, equipment maintenance and wind, and shall provide shielding from solar radiation that can cause degradation of the material. Adhesives tape shall not be permitted.

C403.2.9 Mechanical systems commissioning and completion requirements. Mechanical systems shall be commissioned and completed in accordance with Section C408.2.

C403.2.10 Air system design and control. Each HVAC system having a total fan system motor nameplate horsepower (hp) exceeding 5 horsepower (hp) (3.7 kW) shall meet the provisions of Sections C403.2.10.1 through C403.2.10.2.

C403.2.10.1 Allowable fan motor horsepower. Each HVAC system at fan system design conditions shall not exceed the allowable *fan system motor nameplate hp*

TABLE C403.2.8
MINIMUM PIPE INSULATION THICKNESS (thickness in inches)[a]

FLUID OPERATING TEMPERATURE RANGE AND USAGE (°F)	INSULATION CONDUCTIVITY		NOMINAL PIPE OR TUBE SIZE (inches)				
	Conductivity Btu · in./(h · ft² · °F)[b]	Mean Rating Temperature, °F	< 1	1 to < 1½	1½ to < 4	4 to < 8	≤ 8
> 350	0.32 – 0.34	250	4.5	5.0	5.0	5.0	5.0
251 – 350	0.29 – 0.32	200	3.0	4.0	4.5	4.5	4.5
201 – 250	0.27 – 0.30	150	2.5	2.5	2.5	3.0	3.0
141 – 200	0.25 – 0.29	125	1.5	1.5	2.0	2.0	2.0
105 – 140	0.21 – 0.28	100	1.0	1.0	1.5	1.5	1.5
40 – 60	0.21 – 0.27	75	0.5	0.5	1.0	1.0	1.0
< 40	0.20 – 0.26	75	0.5	1.0	1.0	1.0	1.5

a. For piping smaller than 1½ inch (38 mm) and located in partitions within *conditioned spaces*, reduction of these thicknesses by 1 inch (25 mm) shall be permitted (before thickness adjustment required in footnote b) but not to a thickness less than 1 inch (25 mm).

b. For insulation outside the stated conductivity range, the minimum thickness (*T*) shall be determined as follows:

$T = r\{(1 + t/r)^{K/k} - 1\}$

where:

T = minimum insulation thickness,

r = actual outside radius of pipe,

t = insulation thickness listed in the table for applicable fluid temperature and pipe size,

K = conductivity of alternate material at mean rating temperature indicated for the applicable fluid temperature (Btu × in/h × ft² × °F) and

k = the upper value of the conductivity range listed in the table for the applicable fluid temperature.

c. For direct-buried heating and hot water system piping, reduction of these thicknesses by 1½ inches (38 mm) shall be permitted (before thickness adjustment required in footnote b but not to thicknesses less than 1 inch (25 mm).

(Option 1) or *fan system bhp* (Option 2) as shown in Table C403.2.10.1(1). This includes supply fans, return/relief fans, and fan-powered terminal units associated with systems providing heating or cooling capability. Single *zone* variable-air-volume systems shall comply with the constant volume fan power limitation.

Exception: The following fan systems are exempt from allowable fan floor horsepower requirement.

1. Hospital, vivarium and laboratory systems that utilize flow control devices on exhaust and/or return to maintain space pressure relationships necessary for occupant health and safety or environmental control shall be permitted to use variable volume fan power limitation.

2. Individual exhaust fans with motor nameplate horsepower of 1 hp or less.

C403.2.10.2 Motor nameplate horsepower. For each fan, the selected fan motor shall be no larger than the first available motor size greater than the brake horsepower

(bhp). The fan brake horsepower (bhp) shall be indicated on the design documents to allow for compliance verification by the *code official*.

Exceptions:

1. For fans less than 6 bhp (4413 W), where the first available motor larger than the brake horsepower has a nameplate rating within 50 percent of the bhp, selection of the next larger nameplate motor size is allowed.

2. For fans 6 bhp (4413 W) and larger, where the first available motor larger than the bhp has a nameplate rating within 30 percent of the bhp, selection of the next larger nameplate motor size is allowed.

C403.2.11 Heating outside a building. Systems installed to provide heat outside a building shall be radiant systems.

Such heating systems shall be controlled by an occupancy sensing device or a timer switch, so that the system is automatically deenergized when no occupants are present.

TABLE C403.2.10.1(1)
FAN POWER LIMITATION

	LIMIT	CONSTANT VOLUME	VARIABLE VOLUME
Option 1: Fan system motor nameplate hp	Allowable nameplate motor hp	$hp \leq CFM_S \times 0.0011$	$hp \leq CFM_S \times 0.0015$
Option 2: Fan system bhp	Allowable fan system bhp	$bhp \leq CFM_S \times 0.00094 + A$	$bhp \leq CFM_S \times 0.0013 + A$

where:

CFM_S = The maximum design supply airflow rate to conditioned spaces served by the system in cubic feet per minute.
hp = The maximum combined motor nameplate horsepower.
Bhp = The maximum combined fan brake horsepower.
A = Sum of $[PD \times CFM_D / 4131]$

For SI: 1 cfm = 0.471 L/s.

where:

PD = Each applicable pressure drop adjustment from Table C403.2.10.1(2) in. w.c.
CFM_D = The design airflow through each applicable device from Table C403.2.10.1(2) in cubic feet per minute.

For SI: 1 bhp = 735.5 W, 1 hp = 745.5 W.

TABLE C403.2.10.1(2)
FAN POWER LIMITATION PRESSURE DROP ADJUSTMENT

DEVICE	ADJUSTMENT
Credits	
Fully ducted return and/or exhaust air systems	0.5 inch w.c. (2.15 in w.c. for laboratory and vivarium systems)
Return and/or exhaust air flow control devices	0.5 inch w.c.
Exhaust filters, scrubbers, or other exhaust treatment.	The pressure drop of device calculated at fan system design condition
Particulate filtration credit: MERV 9 thru 12	0.5 inch w.c.
Particulate filtration credit: MERV 13 thru 15	0.9 inch. w.c.
Particulate filtration credit: MERV 16 and greater and electronically enhanced filters	Pressure drop calculated at 2x clean filter pressure drop at fan system design condition.
Carbon and other gas-phase air cleaners	Clean filter pressure drop at fan system design condition.
Biosafety cabinet	Pressure drop of device at fan system design condition.
Energy recovery device, other than coil runaround loop	(2.2 × energy recovery effectiveness) – 0.5 inch w.c. for each airstream
Coil runaround loop	0.6 inch w.c. for each airstream
Evaporative humidifier/cooler in series with another cooling coil	Pressure drop of device at fan system design conditions
Sound attenuation section	0.15 inch w.c.
Exhaust system serving fume hoods	0.35 inch w.c.
Laboratory and vivarium exhaust systems in high-rise buildings	0.25 inch w.c./100 feet of vertical duct exceeding 75 feet

w.c. = water column
For SI:1 inch w.c. = 249 Pa, 1 inch = 25.4 mm.

C403.3 Simple HVAC systems and equipment (Prescriptive). This section applies to buildings served by unitary or packaged HVAC equipment listed in Tables C403.2.3(1) through C403.2.3(8), each serving one *zone* and controlled by a single thermostat in the *zone* served. It also applies to two-pipe heating systems serving one or more *zones*, where no cooling system is installed.

C403.3.1 Economizers. Each cooling system that has a fan shall include either an air or water economizer meeting the requirements of Sections C403.3.1.1 through C403.3.1.1.4.

Exception: Economizers are not required for the systems listed below.

1. Individual fan-cooling units with a supply capacity less than the minimum listed in Table C403.3.1(1).

2. Where more than 25 percent of the air designed to be supplied by the system is to spaces that are designed to be humidified above 35°F (1.7 °C) dew-point temperature to satisfy process needs.

3. Systems that serve *residential* spaces where the system capacity is less than five times the requirement listed in Table C403.3.1(1).

4. Systems expected to operate less than 20 hours per week.

5. Where the use of *outdoor air* for cooling will affect supermarket open refrigerated casework systems.

6. Where the cooling *efficiency* meets or exceeds the *efficiency* requirements in Table C403.3.1(2).

TABLE C403.3.1(1)
ECONOMIZER REQUIREMENTS

CLIMATE ZONES	ECONOMIZER REQUIREMENT
1A, 1B	No requirement
2A, 2B, 3A, 3B, 3C, 4A, 4B, 4C, 5A, 5B, 5C, 6A, 6B, 7, 8	Economizers on all cooling systems ≥ 33,000 Btu/h[a]

For SI:1 British thermal unit per hour = 0.2931 W.

a. The total capacity of all systems without economizers shall not exceed 300,000 Btu/h per *building*, or 20 percent of its air economizer capacity, whichever is greater.

TABLE C403.3.1(2)
EQUIPMENT EFFICIENCY PERFORMANCE EXCEPTION FOR ECONOMIZERS

CLIMATE ZONES	COOLING EQUIPMENT PERFORMANCE IMPROVEMENT (EER OR IPLV)
2B	10% Efficiency Improvement
3B	15% Efficiency Improvement
4B	20% Efficiency Improvement

C403.3.1.1 Air economizers. Air economizers shall comply with Sections C403.3.1.1.1 through C403.3.1.1.4.

C403.3.1.1.1 Design capacity. Air economizer systems shall be capable of modulating *outdoor air* and return air dampers to provide up to 100 percent of the design supply air quantity as *outdoor air* for cooling.

C403.3.1.1.2 Control signal. Economizer dampers shall be capable of being sequenced with the mechanical cooling equipment and shall not be controlled by only mixed air temperature.

Exception: The use of mixed air temperature limit control shall be permitted for systems controlled from space temperature (such as single-*zone* systems).

C403.3.1.1.3 High-limit shutoff. Air economizers shall be capable of automatically reducing *outdoor air* intake to the design minimum *outdoor air* quantity when *outdoor air* intake will no longer reduce cooling energy usage. High-limit shutoff control types for specific climates shall be chosen from Table C403.3.1.1.3(1). High-limit shutoff control settings for these control types shall be those specified in Table C403.3.1.1.3(2).

C403.3.1.1.4 Relief of excess outdoor air. Systems shall be capable of relieving excess *outdoor air* during air economizer operation to prevent over-pressurizing the building. The relief air outlet shall be located to avoid recirculation into the building.

C403.3.2 Hydronic system controls. Hydronic systems of at least 300,000 Btu/h (87 930 W) design output capacity supplying heated and chilled water to comfort conditioning systems shall include controls that meet the requirements of Section C403.4.3.

C403.4 Complex HVAC systems and equipment. (Prescriptive). This section applies to buildings served by HVAC equipment and systems not covered in Section C403.3.

C403.4.1 Economizers. Economizers shall comply with Sections C403.4.1.1 through C403.4.1.4.

C403.4.1.1 Design capacity. Water economizer systems shall be capable of cooling supply air by indirect evaporation and providing up to 100 percent of the expected system cooling load at *outdoor air* temperatures of 50°F dry bulb (10°C dry bulb)/45°F wet bulb (7.2°C wet bulb) and below.

Exception: Systems in which a water economizer is used and where dehumidification requirements cannot be met using outdoor air temperatures of 50°F dry bulb (10°C dry bulb)/45°F wet bulb (7.2°C wet bulb) shall satisfy 100 percent of the expected system cooling load at 45°F dry bulb (7.2°C dry bulb)/ 40°F wet bulb (4.5°C wet bulb).

C403.4.1.2 Maximum pressure drop. Precooling coils and water-to-water heat exchangers used as part of a water economizer system shall either have a water-side pressure drop of less than 15 feet (4572 mm) of water or a secondary loop shall be created so that the coil or heat exchanger pressure drop is not seen by the circulating pumps when the system is in the normal cooling (noneconomizer) mode.

TABLE C403.3.1.1.3(1)
HIGH-LIMIT SHUTOFF CONTROL OPTIONS FOR AIR ECONOMIZERS

CLIMATE ZONES	ALLOWED CONTROL TYPES	PROHIBITED CONTROL TYPES
1B, 2B, 3B, 3C, 4B, 4C, 5B, 5C, 6B, 7, 8	Fixed dry bulb Differential dry bulb Electronic enthalpy[a] Differential enthalpy Dew-point and dry-bulb temperatures	Fixed enthalpy
1A, 2A, 3A, 4A	Fixed dry bulb Fixed enthalpy Electronic enthalpy[a] Differential enthalpy Dew-point and dry-bulb temperatures	Differential dry bulb
All other climates	Fixed dry bulb Differential dry bulb Fixcd enthalpy Electronic enthalpy[a] Differential enthalpy Dew-point and dry-bulb temperatures	—

a. Electronic enthalpy controllers are devices that use a combination of humidity and dry-bulb temperature in their switching algorithm.

TABLE C403.3.1.1.3(2)
HIGH-LIMIT SHUTOFF CONTROL SETTING FOR AIR ECONOMIZERS

DEVICE TYPE	CLIMATE ZONE	REQUIRED HIGH LIMIT (ECONOMIZER OFF WHEN):	
		EQUATION	DESCRIPTION
Fixed dry bulb	1B, 2B, 3B, 3C, 4B, 4C, 5B, 5C, 6B, 7, 8	$T_{OA} > 75°F$	Outdoor air temperature exceeds 75°F
	5A, 6A, 7A	$T_{OA} > 70°F$	Outdoor air temperature exceeds 70°F
	All other zones	$T_{OA} > 65°F$	Outdoor air temperature exceeds 65°F
Differential dry bulb	1B, 2B, 3B, 3C, 4B, 4C, 5A, 5B, 5C, 6A, 6B, 7, 8	$T_{OA} > T_{RA}$	Outdoor air temperature exceeds return air temperature
Fixed enthalpy	All	$h_{OA} > 28$ Btu/lb[a]	Outdoor air enthalpy exceeds 28 Btu/lb of dry air[a]
Electronic Enthalpy	All	$(T_{OA}, RH_{OA}) > A$	Outdoor air temperature/RH exceeds the "A" setpoint curve[b]
Differential enthalpy	All	$h_{OA} > h_{RA}$	Outdoor air enthalpy exceeds return air enthalpy
Dew-point and dry bulb temperatures	All	$DP_{OA} > 55°F$ or $T_{OA} > 75°F$	Outdoor air dry bulb exceeds 75°F or outside dew point exceeds 55°F (65 gr/lb)

For SI: °C = (°F - 32) × $^5/_9$, 1 Btu/lb = 2.33 kJ/kg.

a. At altitudes substantially different than sea level, the Fixed Enthalpy limit shall be set to the enthalpy value at 75°F and 50-percent relative humidity. As an example, at approximately 6,000 feet elevation the fixed enthalpy limit is approximately 30.7 Btu/lb.

b. Setpoint "A" corresponds to a curve on the psychometric chart that goes through a point at approximately 75°F and 40-percent relative humidity and is nearly parallel to dry-bulb lines at low humidity levels and nearly parallel to enthalpy lines at high humidity levels.

C403.4.1.3 Integrated economizer control. Economizer systems shall be integrated with the mechanical cooling system and be capable of providing partial cooling even where additional mechanical cooling is required to meet the remainder of the cooling load.

Exceptions:

1. Direct expansion systems that include controls that reduce the quantity of *outdoor air* required to prevent coil frosting at the lowest step of compressor unloading, provided this lowest step is no greater than 25 percent of the total system capacity.

2. Individual direct expansion units that have a rated cooling capacity less than 54,000 Btu/h (15 827 W) and use nonintegrated economizer controls that preclude simultaneous operation of the economizer and mechanical cooling.

C403.4.1.4 Economizer heating system impact. HVAC system design and economizer controls shall be such that economizer operation does not increase the building heating energy use during normal operation.

Exception: Economizers on VAV systems that cause *zone* level heating to increase due to a reduction in supply air temperature.

C403.4.2 Variable air volume (VAV) fan control. Individual VAV fans with motors of 7.5 horsepower (5.6 kW) or greater shall be:

1. Driven by a mechanical or electrical variable speed drive;

2. Driven by a vane-axial fan with variable-pitch blades; or

3. The fan shall have controls or devices that will result in fan motor demand of no more than 30 percent of their design wattage at 50 percent of design airflow when static pressure set point equals one-third of the total design static pressure, based on manufacturer's certified fan data.

C403.4.2.1 Static pressure sensor location. Static pressure sensors used to control VAV fans shall be placed in a position such that the controller setpoint is no greater than one-third the total design fan static pressure, except for systems with *zone* reset control complying with Section C403.4.2.2. For sensors installed down-stream of major duct splits, at least one sensor shall be located on each major branch to ensure that static pressure can be maintained in each branch.

C403.4.2.2 Set points for direct digital control. For systems with direct digital control of individual *zone* boxes reporting to the central control panel, the static pressure set point shall be reset based on the *zone* requiring the most pressure, i.e., the set point is reset lower until one *zone* damper is nearly wide open.

C403.4.3 Hydronic systems controls. The heating of fluids that have been previously mechanically cooled and the cooling of fluids that have been previously mechanically heated shall be limited in accordance with Sections

C403.4.3.1 through C403.4.3.3. Hydronic heating systems comprised of multiple-packaged boilers and designed to deliver conditioned water or steam into a common distribution system shall include automatic controls capable of sequencing operation of the boilers. Hydronic heating systems comprised of a single boiler and greater than 500,000 Btu/h (146 550 W) input design capacity shall include either a multistaged or modulating burner.

C403.4.3.1 Three-pipe system. Hydronic systems that use a common return system for both hot water and chilled water are prohibited.

C403.4.3.2 Two-pipe changeover system. Systems that use a common distribution system to supply both heated and chilled water shall be designed to allow a dead band between changeover from one mode to the other of at least 15°F (8.3°C) outside air temperatures; be designed to and provided with controls that will allow operation in one mode for at least 4 hours before changing over to the other mode; and be provided with controls that allow heating and cooling supply temperatures at the changeover point to be no more than 30°F (16.7°C) apart.

C403.4.3.3 Hydronic (water loop) heat pump systems. Hydronic heat pump systems shall comply with Sections C403.4.3.3.1 through C403.4.3.3.3.

C403.4.3.3.1 Temperature dead band. Hydronic heat pumps connected to a common heat pump water loop with central devices for heat rejection and heat addition shall have controls that are capable of providing a heat pump water supply temperature dead band of at least 20°F (11.1°C) between initiation of heat rejection and heat addition by the central devices.

Exception: Where a system loop temperature optimization controller is installed and can determine the most efficient operating temperature based on realtime conditions of demand and capacity, dead bands of less than 20°F (11°C) shall be permitted.

C403.4.3.3.2 Heat rejection. Heat rejection equipment shall comply with Sections C403.4.3.3.2.1 and C403.4.3.3.2.2.

Exception: Where it can be demonstrated that a heat pump system will be required to reject heat throughout the year.

C403.4.3.3.2.1 Climate Zones 3 and 4. For Climate Zones 3 and 4:

1. If a closed-circuit cooling tower is used directly in the heat pump loop, either an automatic valve shall be installed to bypass all but a minimal flow of water around the tower, or lower leakage positive closure dampers shall be provided.

2. If an open-circuit tower is used directly in the heat pump loop, an automatic valve shall be installed to bypass all heat pump water flow around the tower.

3. If an open- or closed-circuit cooling tower is used in conjunction with a separate heat exchanger to isolate the cooling tower from the heat pump loop, then heat loss shall be controlled by shutting down the circulation pump on the cooling tower loop.

C403.4.3.3.2.2 Climate Zones 5 through 8. For Climate Zones 5 through 8, if an open- or closed-circuit cooling tower is used, then a separate heat exchanger shall be provided to isolate the cooling tower from the heat pump loop, and heat loss shall be controlled by shutting down the circulation pump on the cooling tower loop and providing an automatic valve to stop the flow of fluid.

C403.4.3.3.3 Two position valve. Each hydronic heat pump on the hydronic system having a total pump system power exceeding 10 horsepower (hp) (7.5 kW) shall have a two-position valve.

C403.4.3.4 Part load controls. Hydronic systems greater than or equal to 300,000 Btu/h (87 930 W) in design output capacity supplying heated or chilled water to comfort conditioning systems shall include controls that have the capability to:

1. Automatically reset the supply-water temperatures using zone-return water temperature, building-return water temperature, or outside air temperature as an indicator of building heating or cooling demand. The temperature shall be capable of being reset by at least 25 percent of the design supply-to-return water temperature difference; or

2. Reduce system pump flow by at least 50 percent of design flow rate utilizing adjustable speed drive(s) on pump(s), or multiple-staged pumps where at least one-half of the total pump horsepower is capable of being automatically turned off or control valves designed to modulate or step down, and close, as a function of load, or other *approved* means.

C403.4.3.5 Pump isolation. Chilled water plants including more than one chiller shall have the capability to reduce flow automatically through the chiller plant when a chiller is shut down. Chillers piped in series for the purpose of increased temperature differential shall be considered as one chiller.

Boiler plants including more than one boiler shall have the capability to reduce flow automatically through the boiler plant when a boiler is shut down.

C403.4.4 Heat rejection equipment fan speed control. Each fan powered by a motor of 7.5 hp (5.6 kW) or larger shall have the capability to operate that fan at two-thirds of full speed or less, and shall have controls that automatically change the fan speed to control the leaving fluid tem-

perature or condensing temperature/pressure of the heat rejection device.

Exception: Factory-installed heat rejection devices within HVAC equipment tested and rated in accordance with Tables C403.2.3(6) and C403.2.3(7).

C403.4.5 Requirements for complex mechanical systems serving multiple zones. Sections C403.4.5.1 through C403.4.5.4 shall apply to complex mechanical systems serving multiple zones. Supply air systems serving multiple zones shall be VAV systems which, during periods of occupancy, are designed and capable of being controlled to reduce primary air supply to each *zone* to one of the following before reheating, recooling or mixing takes place:

1. Thirty percent of the maximum supply air to each *zone*.

2. Three hundred cfm (142 L/s) or less where the maximum flow rate is less than 10 percent of the total fan system supply airflow rate.

3. The minimum ventilation requirements of Chapter 4 of the *International Mechanical Code*.

Exception: The following define where individual *zones* or where entire air distribution systems are exempted from the requirement for VAV control:

1. *Zones* where special pressurization relationships or cross-contamination requirements are such that VAV systems are impractical.

2. *Zones* or supply air systems where at least 75 percent of the energy for reheating or for providing warm air in mixing systems is provided from a site-recovered or site-solar energy source.

3. *Zones* where special humidity levels are required to satisfy process needs.

4. *Zones* with a peak supply air quantity of 300 cfm (142 L/s) or less and where the flow rate is less than 10 percent of the total fan system supply airflow rate.

5. *Zones* where the volume of air to be reheated, recooled or mixed is no greater than the volume of outside air required to meet the minimum ventilation requirements of Chapter 4 of the *International Mechanical Code*.

6. *Zones* or supply air systems with thermostatic and humidistatic controls capable of operating in sequence the supply of heating and cooling energy to the *zones* and which are capable of preventing reheating, recooling, mixing or simultaneous supply of air that has been previously cooled, either mechanically or through the use of economizer systems, and air that has been previously mechanically heated.

C403.4.5.1 Single duct variable air volume (VAV) systems, terminal devices. Single duct VAV systems shall use terminal devices capable of reducing the supply of primary supply air before reheating or recooling takes place.

C403.4.5.2 Dual duct and mixing VAV systems, terminal devices. Systems that have one warm air duct and one cool air duct shall use terminal devices which are capable of reducing the flow from one duct to a minimum before mixing of air from the other duct takes place.

C403.4.5.3 Single fan dual duct and mixing VAV systems, economizers. Individual dual duct or mixing heating and cooling systems with a single fan and with total capacities greater than 90,000 Btu/h [(26 375 W) 7.5 tons] shall not be equipped with air economizers.

C403.4.5.4 Supply-air temperature reset controls. Multiple *zone* HVAC systems shall include controls that automatically reset the supply-air temperature in response to representative building loads, or to outdoor air temperature. The controls shall be capable of resetting the supply air temperature at least 25 percent of the difference between the design supply-air temperature and the design room air temperature.

Exceptions:

1. Systems that prevent reheating, recooling or mixing of heated and cooled supply air.

2. Seventy five percent of the energy for reheating is from site-recovered or site solar energy sources.

3. Zones with peak supply air quantities of 300 cfm (142 L/s) or less.

C403.4.6 Heat recovery for service water heating. Condenser heat recovery shall be installed for heating or reheating of service hot water provided the facility operates 24 hours a day, the total installed heat capacity of water-cooled systems exceeds 6,000,000 Btu/hr (1 758 600 W) of heat rejection, and the design service water heating load exceeds 1,000,000 Btu/h (293 100 W).

The required heat recovery system shall have the capacity to provide the smaller of:

1. Sixty percent of the peak heat rejection load at design conditions; or

2. The preheating required to raise the peak service hot water draw to 85°F (29°C).

Exceptions:

1. Facilities that employ condenser heat recovery for space heating or reheat purposes with a heat recovery design exceeding 30 percent of the peak water-cooled condenser load at design conditions.

2. Facilities that provide 60 percent of their service water heating from site solar or site recovered energy or from other sources.

C403.4.7 Hot gas bypass limitation. Cooling systems shall not use hot gas bypass or other evaporator pressure control systems unless the system is designed with multiple steps of unloading or continuous capacity modulation. The capacity of the hot gas bypass shall be limited as indicated in Table C403.4.7

Exception: Unitary packaged systems with cooling capacities not greater than 90,000 Btu/h (26 379 W).

TABLE C403.4.7
MAXIMUM HOT GAS BYPASS CAPACITY

RATED CAPACITY	MAXIMUM HOT GAS BYPASS CAPACITY (% of total capacity)
≤ 240,000 Btu/h	50
> 240,000 Btu/h	25

For SI: 1 British thermal unit per hour = 0.2931 W.

SECTION C404
SERVICE WATER HEATING
(Mandatory)

C404.1 General. This section covers the minimum efficiency of, and controls for, service water-heating equipment and insulation of service hot water piping.

C404.2 Service water-heating equipment performance efficiency. Water-heating equipment and hot water storage tanks shall meet the requirements of Table C404.2. The efficiency shall be verified through data furnished by the manufacturer or through certification under an *approved* certification program.

C404.3 Temperature controls. Service water-heating equipment shall be provided with controls to allow a setpoint of 110°F (43°C) for equipment serving dwelling units and 90°F (32°C) for equipment serving other occupancies. The outlet temperature of lavatories in public facility rest rooms shall be limited to 110°F (43°C).

C404.4 Heat traps. Water-heating equipment not supplied with integral heat traps and serving noncirculating systems shall be provided with heat traps on the supply and discharge piping associated with the equipment.

C404.5 Pipe insulation. For automatic-circulating hot water and heat-traced systems, piping shall be insulated with not less than 1 inch (25 mm) of insulation having a conductivity not exceeding 0.27 Btu per inch/h × ft² × °F (1.53 W per 25 mm/ m² × K). The first 8 feet (2438 mm) of piping in non-hot-water-supply temperature maintenance systems served by equipment without integral heat traps shall be insulated with 0.5 inch (12.7 mm) of material having a conductivity not exceeding 0.27 Btu per inch/h × ft² × °F (1.53 W per 25 mm/m² × K).

Exception: Heat-traced piping systems shall meet the insulation thickness requirements per the manufacturer's installation instructions. Untraced piping within a heat traced system shall be insulated with not less than 1 inch (25 mm) of insulation having a conductivity not exceeding 0.27 Btu per inch/h × ft² × °F (1.53 W per 25 mm/m² × K).

TABLE C404.2
MINIMUM PERFORMANCE OF WATER-HEATING EQUIPMENT

EQUIPMENT TYPE	SIZE CATEGORY (input)	SUBCATEGORY OR RATING CONDITION	PERFORMANCE REQUIRED[a, b]	TEST PROCEDURE
Water heaters, electric	≤ 12 kW	Resistance	0.97 - 0.00 132V, EF	DOE 10 CFR Part 430
	> 12 kW	Resistance	1.73V+ 155 SL, Btu/h	ANSI Z21.10.3
	≤ 24 amps and ≤ 250 volts	Heat pump	0.93 - 0.00 132V, EF	DOE 10 CFR Part 430
Storage water heaters, gas	≤ 75,000 Btu/h	≥ 20 gal	0.67 - 0.0019V, EF	DOE 10 CFR Part 430
	> 75,000 Btu/h and ≤ 155,000 Btu/h	< 4,000 Btu/h/gal	80% E_t (Q/800 + 110$\sqrt{V}$)SL, Btu/h	ANSI Z21.10.3
	> 155,000 Btu/h	< 4,000 Btu/h/gal	80% E_t (Q/800 + 110$\sqrt{V}$)SL, Btu/h	
Instantaneous water heaters, gas	> 50,000 Btu/h and < 200,000 Btu/h[c]	≥ 4,000 (Btu/h)/gal and < 2 gal	0.62 - 0.00 19V, EF	DOE 10 CFR Part 430
	≥ 200,000 Btu/h	≥ 4,000 Btu/h/gal and < 10 gal	80% E_t	ANSI Z21.10.3
	≥ 200,000 Btu/h	≥ 4,000 Btu/h/gal and ≥ 10 gal	80% E_t (Q/800 + 110$\sqrt{V}$)SL, Btu/h	
Storage water heaters, oil	≤ 105,000 Btu/h	≥ 20 gal	0.59 - 0.0019V, EF	DOE 10 CFR Part 430
	≥ 105,000 Btu/h	< 4,000 Btu/h/gal	78% E_t (Q/800 + 110$\sqrt{V}$)SL, Btu/h	ANSI Z21.10.3
Instantaneous water heaters, oil	≤ 210,000 Btu/h	≥ 4,000 Btu/h/gal and < 2 gal	0.59 - 0.0019V, EF	DOE 10 CFR Part 430
	> 210,000 Btu/h	≥ 4,000 Btu/h/gal and < 10 gal	80% E_t	ANSI Z21.10.3
	> 210,000 Btu/h	≥ 4,000 Btu/h/gal and ≥ 10 gal	78% E_t (Q/800 + 110$\sqrt{V}$)SL, Btu/h	
Hot water supply boilers, gas and oil	≥ 300,000 Btu/h and < 12,500,000 Btu/h	≥ 4,000 Btu/h/gal and < 10 gal	80% E_t	ANSI Z21.10.3
Hot water supply boilers, gas	≥ 300,000 Btu/h and < 12,500,000 Btu/h	≥ 4,000 Btu/h/gal and ≥ 10 gal	80% E_t (Q/800 + 110$\sqrt{V}$)SL, Btu/h	
Hot water supply boilers, oil	> 300,000 Btu/h and < 12,500,000 Btu/h	> 4,000 Btu/h/gal and > 10 gal	78% E_t (Q/800 + 110$\sqrt{V}$)SL, Btu/h	
Pool heaters, gas and oil	All	—	78% E_t	ASHRAE 146
Heat pump pool heaters	All	—	4.0 COP	AHRI 1160
Unfired storage tanks	All	—	Minimum insulation requirement R-12.5 (h · ft² · °F)/Btu	(none)

For SI: °C = [(°F) - 32]/1.8, 1 British thermal unit per hour = 0.2931 W, 1 gallon = 3.785 L, 1 British thermal unit per hour per gallon = 0.078 W/L.

a. Energy factor (EF) and thermal efficiency (E_t) are minimum requirements. In the EF equation, V is the rated volume in gallons.

b. Standby loss (SL) is the maximum Btu/h based on a nominal 70°F temperature difference between stored water and ambient requirements. In the SL equation, Q is the nameplate input rate in Btu/h. In the SL equation for electric water heaters, V is the rated volume in gallons. In the SL equation for oil and gas water heaters and boilers, V is the rated volume in gallons.

c. Instantaneous water heaters with input rates below 200,000 Btu/h must comply with these requirements if the water heater is designed to heat water to temperatures 180°F or higher.

C404.6 Hot water system controls. Circulating hot water system pumps or heat trace shall be arranged to be turned off either automatically or manually when there is limited hot water demand. Ready access shall be provided to the operating controls.

C404.7 Pools and inground permanently installed spas (Mandatory). Pools and inground permanently installed spas shall comply with Sections C404.7.1 through C404.7.3.

C404.7.1 Heaters. All heaters shall be equipped with a readily *accessible* on-off switch that is mounted outside of the heater to allow shutting off the heater without adjusting the thermostat setting. Gas-fired heaters shall not be equipped with constant burning pilot lights.

C404.7.2 Time switches. Time switches or other control method that can automatically turn off and on heaters and pumps according to a preset schedule shall be installed on all heaters and pumps. Heaters, pumps and motors that have built in timers shall be deemed in compliance with this requirement.

> **Exceptions:**
> 1. Where public health standards require 24-hour pump operation.
> 2. Where pumps are required to operate solar- and waste-heat-recovery pool heating systems.

C404.7.3 Covers. Heated pools and inground permanently installed spas shall be provided with a vapor-retardant cover.

> **Exception:** A vapor-retardant cover is not required for pools deriving over 70 percent of the energy for heating from site-recovered energy, such as a heat pump or solar energy source computed over an operating season.

SECTION C405
ELECTRICAL POWER AND LIGHTING SYSTEMS
(MANDATORY)

C405.1 General (Mandatory). This section covers lighting system controls, the connection of ballasts, the maximum lighting power for interior applications, electrical energy consumption, and minimum acceptable lighting equipment for exterior applications.

> **Exception:** Dwelling units within commercial buildings shall not be required to comply with Sections C405.2 through C405.5 provided that not less than 75 percent of the permanently installed light fixtures, other than low-voltage lighting, shall be fitted for, and contain only, high efficacy lamps.

C405.2 Lighting controls (Mandatory). Lighting systems shall be provided with controls as specified in Sections C405.2.1, C405.2.2, C405.2.3 and C405.2.4.

C405.2.1 Manual lighting controls. All buildings shall include manual lighting controls that meet the requirements of Sections C405.2.1.1 and C405.2.1.2.

C405.2.1.1 Interior lighting controls. Each area enclosed by walls or floor-to-ceiling partitions shall have at least one manual control for the lighting serving that area. The required controls shall be located within the area served by the controls or be a remote switch that identifies the lights served and indicates their status.

> **Exceptions:**
> 1. Areas designated as security or emergency areas that need to be continuously lighted.
> 2. Lighting in stairways or corridors that are elements of the means of egress.

C405.2.1.2 Light reduction controls. Each area that is required to have a manual control shall also allow the occupant to reduce the connected lighting load in a reasonably uniform illumination pattern by at least 50 percent. Lighting reduction shall be achieved by one of the following or other *approved* method:

1. Controlling all lamps or luminaires;
2. Dual switching of alternate rows of luminaires, alternate luminaires or alternate lamps;
3. Switching the middle lamp luminaires independently of the outer lamps; or
4. Switching each luminaire or each lamp.

> **Exception:** Light reduction controls need not be provided in the following areas and spaces:
> 1. Areas that have only one luminaire, with rated power less than 100 watts.
> 2. Areas that are controlled by an occupant-sensing device.
> 3. Corridors, equipment rooms, storerooms, restrooms, public lobbies, electrical or mechanical rooms.
> 4. *Sleeping unit* (see Section C405.2.3).
> 5. Spaces that use less than 0.6 watts per square foot (6.5 W/m^2).
> 6. Daylight spaces complying with Section C405.2.2.3.2.

C405.2.2 Additional lighting controls. Each area that is required to have a manual control shall also have controls that meet the requirements of Sections C405.2.2.1, C405.2.2.2 and C405.2.2.3.

> **Exception:** Additional lighting controls need not be provided in the following spaces:
> 1. *Sleeping units*.
> 2. Spaces where patient care is directly provided.
> 3. Spaces where an automatic shutoff would endanger occupant safety or security.
> 4. Lighting intended for continuous operation.

C405.2.2.1 Automatic time switch control devices. Automatic time switch controls shall be installed to control lighting in all areas of the building.

> **Exceptions:**
> 1. Emergency egress lighting does not need to be controlled by an automatic time switch.

2. Lighting in spaces controlled by occupancy sensors does not need to be controlled by automatic time switch controls.

The automatic time switch control device shall include an override switching device that complies with the following:

1. The override switch shall be in a readily accessible location;

2. The override switch shall be located where the lights controlled by the switch are visible; or the switch shall provide a mechanism which announces the area controlled by the switch;

3. The override switch shall permit manual operation;

4. The override switch, when initiated, shall permit the controlled lighting to remain on for a maximum of 2 hours; and

5. Any individual override switch shall control the lighting for a maximum area of 5,000 square feet (465 m²).

Exception: Within malls, arcades, auditoriums, single tenant retail spaces, industrial facilities and arenas:

1. The time limit shall be permitted to exceed 2 hours provided the override switch is a captive key device; and

2. The area controlled by the override switch is permitted to exceed 5,000 square feet (465 m²), but shall not exceed 20,000 square feet (1860 m²).

C405.2.2.2 Occupancy sensors. Occupancy sensors shall be installed in all classrooms, conference/meeting rooms, employee lunch and break rooms, private offices, restrooms, storage rooms and janitorial closets, and other spaces 300 square feet (28 m²) or less enclosed by floor-to-ceiling height partitions. These automatic control devices shall be installed to automatically turn off lights within 30 minutes of all occupants leaving the space, and shall either be manual on or shall be controlled to automatically turn the lighting on to not more than 50 percent power.

Exception: Full automatic-on controls shall be permitted to control lighting in public corridors, stairways, restrooms, primary building entrance areas and lobbies, and areas where manual-on operation would endanger the safety or security of the room or building occupants

C405.2.2.3 Daylight zone control. Daylight zones shall be designed such that lights in the daylight zone are controlled independently of general area lighting and are controlled in accordance with either Section C405.2.2.3.1 or Section C405.2.2.3.2. Each daylight control zone shall not exceed 2,500 square feet (232 m²). Contiguous daylight zones adjacent to vertical fenestration are allowed to be controlled by a single con-

trolling device provided that they do not include zones facing more than two adjacent cardinal orientations (i.e., north, east, south, west). Daylight zones under skylights more than 15 feet (4572 mm) from the perimeter shall be controlled separately from daylight zones adjacent to vertical fenestration.

Exception: Daylight zones enclosed by walls or ceiling height partitions and containing two or fewer light fixtures are not required to have a separate switch for general area lighting.

C405.2.2.3.1 Manual daylighting controls. Manual controls shall be installed in daylight zones unless automatic controls are installed in accordance with Section C405.2.2.3.2.

C405.2.2.3.2 Automatic daylighting controls. Set-point and other controls for calibrating the lighting control device shall be readily accessible.

Daylighting controls device shall be capable of automatically reducing the lighting power in response to available daylight by either one of the following methods:

1. Continuous dimming using dimming ballasts and daylight-sensing automatic controls that are capable of reducing the power of general lighting in the daylit zone continuously to less than 35 percent of rated power at maximum light output.

2. Stepped dimming using multi-level switching and daylight-sensing controls that are capable of reducing lighting power automatically. The system shall provide a minimum of two control channels per zone and be installed in a manner such that at least one control step is between 50 percent and 70 percent of design lighting power and another control step is no greater than 35 percent of design power.

C405.2.2.3.3 Multi-level lighting controls. Where multi-level lighting controls are required by this code, the general lighting in the daylight zone shall be separately controlled by at least one multi-level lighting control that reduces the lighting power in response to daylight available in the space. Where the daylit illuminance in the space is greater than the rated illuminance of the general lighting of daylight zones, the general lighting shall be automatically controlled so that its power draw is no greater than 35 percent of its rated power. The multi-level lighting control shall be located so that calibration and set point adjustment controls are readily accessible and separate from the light sensor.

C405.2.3 Specific application controls. Specific application controls shall be provided for the following:

1. Display and accent light shall be controlled by a dedicated control which is independent of the controls for other lighting within the room or space.

2. Lighting in cases used for display case purposes shall be controlled by a dedicated control which is independent of the controls for other lighting within the room or space.

3. Hotel and motel sleeping units and guest suites shall have a master control device at the main room entry that controls all permanently installed luminaires and switched receptacles.

4. Supplemental task lighting, including permanently installed under-shelf or under-cabinet lighting, shall have a control device integral to the luminaires or be controlled by a wall-mounted control device provided the control device is readily accessible.

5. Lighting for nonvisual applications, such as plant growth and food warming, shall be controlled by a dedicated control which is independent of the controls for other lighting within the room or space.

6. Lighting equipment that is for sale or for demonstrations in lighting education shall be controlled by a dedicated control which is independent of the controls for other lighting within the room or space.

C405.2.4 Exterior lighting controls. Lighting not designated for dusk-to-dawn operation shall be controlled by either a combination of a photosensor and a time switch, or an astronomical time switch. Lighting designated for dusk-to-dawn operation shall be controlled by an astronomical time switch or photosensor. All time switches shall be capable of retaining programming and the time setting during loss of power for a period of at least 10 hours.

C405.3 Tandem wiring (Mandatory). The following luminaires located within the same area shall be tandem wired:

1. Fluorescent luminaires equipped with one, three or odd-numbered lamp configurations, that are recess-mounted within 10 feet (3048 mm) center-to-center of each other.

2. Fluorescent luminaires equipped with one, three or any odd-numbered lamp configuration, that are pendant- or surface-mounted within 1 foot (305 mm) edge-to-edge of each other.

Exceptions:

1. Where electronic high-frequency ballasts are used.

2. Luminaires on emergency circuits.

3. Luminaires with no available pair in the same area.

C405.4 Exit signs (Mandatory). Internally illuminated exit signs shall not exceed 5 watts per side.

C405.5 Interior lighting power requirements (Prescriptive). A building complies with this section if its total connected lighting power calculated under Section C405.5.1 is no greater than the interior lighting power calculated under Section C405.5.2.

C405.5.1 Total connected interior lighting power. The total connected interior lighting power (watts) shall be the sum of the watts of all interior lighting equipment as deter-

mined in accordance with Sections C405.5.1.1 through C405.5.1.4.

Exceptions:

1. The connected power associated with the following lighting equipment is not included in calculating total connected lighting power.

    1.1. Professional sports arena playing field lighting.

    1.2. *Sleeping unit* lighting in hotels, motels, boarding houses or similar buildings.

    1.3. Emergency lighting automatically off during normal building operation.

    1.4. Lighting in spaces specifically designed for use by occupants with special lighting needs including the visually impaired visual impairment and other medical and age-related issues.

    1.5. Lighting in interior spaces that have been specifically designated as a registered interior historic landmark.

    1.6. Casino gaming areas.

2. Lighting equipment used for the following shall be exempt provided that it is in addition to general lighting and is controlled by an independent control device:

    2.1. Task lighting for medical and dental purposes.

    2.2. Display lighting for exhibits in galleries, museums and monuments.

3. Lighting for theatrical purposes, including performance, stage, film production and video production.

4. Lighting for photographic processes.

5. Lighting integral to equipment or instrumentation and is installed by the manufacturer.

6. Task lighting for plant growth or maintenance.

7. Advertising signage or directional signage.

8. In restaurant buildings and areas, lighting for food warming or integral to food preparation equipment.

9. Lighting equipment that is for sale.

10. Lighting demonstration equipment in lighting education facilities.

11. Lighting *approved* because of safety or emergency considerations, inclusive of exit lights.

12. Lighting integral to both open and glass-enclosed refrigerator and freezer cases.

13. Lighting in retail display windows, provided the display area is enclosed by ceiling-height partitions.

14. Furniture mounted supplemental task lighting that is controlled by automatic shutoff.

C405.5.1.1 Screw lamp holders. The wattage shall be the maximum *labeled* wattage of the luminaire.

C405.5.1.2 Low-voltage lighting. The wattage shall be the specified wattage of the transformer supplying the system.

C405.5.1.3 Other luminaires. The wattage of all other lighting equipment shall be the wattage of the lighting equipment verified through data furnished by the manufacturer or other *approved* sources.

C405.5.1.4 Line-voltage lighting track and plug-in busway. The wattage shall be:

1. The specified wattage of the luminaires included in the system with a minimum of 30 W/lin ft. (98 W/lin. m);

2. The wattage limit of the system's circuit breaker; or

3. The wattage limit of other permanent current limiting device(s) on the system.

C405.5.2 Interior lighting power. The total interior lighting power allowance (watts) is determined according to Table C405.5.2(1) using the Building Area Method, or Table C405.5.2(2) using the Space-by-Space Method, for all areas of the building covered in this permit. For the Building Area Method, the interior lighting power allowance is the floor area for each building area type listed in Table C405.5.2(1) times the value from Table C405.5.2(1) for that area. For the purposes of this method, an "area" shall be defined as all contiguous spaces that accommodate or are associated with a single building area type as listed in Table C405.5.2(1). Where this method is used to calculate the total interior lighting power for an entire building, each building area type shall be treated as a separate area. For the Space-by-Space Method, the interior lighting power allowance is determined by multiplying the floor area of each space times the value for the space type in Table C405.5.2(2) that most closely represents the proposed use of the space, and then summing the lighting power allowances for all spaces. Tradeoffs among spaces are permitted.

C405.6 Exterior lighting (Mandatory). Where the power for exterior lighting is supplied through the energy service to the building, all exterior lighting, other than low-voltage landscape lighting, shall comply with Sections C405.6.1 and C405.6.2.

> **Exception:** Where *approved* because of historical, safety, signage or emergency considerations.

C405.6.1 Exterior building grounds lighting. All exterior building grounds luminaires that operate at greater than 100 watts shall contain lamps having a minimum efficacy of 60 lumens per watt unless the luminaire is controlled by a motion sensor or qualifies for one of the exceptions under Section C405.6.2.

TABLE C405.5.2(1)
INTERIOR LIGHTING POWER ALLOWANCES:
BUILDING AREA METHOD

BUILDING AREA TYPE	LPD (w/ft²)
Automotive facility	0.9
Convention center	1.2
Courthouse	1.2
Dining: bar lounge/leisure	1.3
Dining: cafeteria/fast food	1.4
Dining: family	1.6
Dormitory	1.0
Exercise center	1.0
Fire station	0.8
Gymnasium	1.1
Health care clinic	1.0
Hospital	1.2
Hotel	1.0
Library	1.3
Manufacturing facility	1.3
Motel	1.0
Motion picture theater	1.2
Multifamily	0.7
Museum	1.1
Office	0.9
Parking garage	0.3
Penetentiary	1.0
Performing arts theater	1.6
Police station	1.0
Post office	1.1
Religious building	1.3
Retail	1.4
School/university	1.2
Sports arena	1.1
Town hall	1.1
Transportation	1.0
Warehouse	0.6
Workshop	1.4

TABLE C405.5.2(2)
INTERIOR LIGHTING POWER ALLOWANCES:
SPACE-BY-SPACE METHOD

COMMON SPACE-BY-SPACE TYPES	LPD (w/ft²)
Atrium – First 40 feet in height	0.03 per ft. ht.
Atrium – Above 40 feet in height	0.02 per ft. ht.
Audience/seating area – permanent	
For auditorium	0.9
For performing arts theater	2.6
For motion picture theater	1.2
Classroom/lecture/training	1.30
Conference/meeting/multipurpose	1.2
Corridor/transition	0.7
Dining area	
Bar/lounge/leisure dining	1.40
Family dining area	1.40
Dressing/fitting room performing arts theater	1.1
Electrical/mechanical	1.10
Food preparation	1.20
Laboratory for classrooms	1.3
Laboratory for medical/industrial/research	1.8
Lobby	1.10
Lobby for performing arts theater	3.3
Lobby for motion picture theater	1.0
Locker room	0.80
Lounge recreation	0.8
Office – enclosed	1.1
Office – open plan	1.0
Restroom	1.0
Sales area	1.6[a]
Stairway	0.70
Storage	0.8
Workshop	1.60
Courthouse/police station/penetentiary	
Courtroom	1.90
Confinement cells	1.1
Judge chambers	1.30
Penitentiary audience seating	0.5
Penitentiary classroom	1.3
Penitentiary dining	1.1
BUILDING SPECIFIC SPACE-BY-SPACE TYPES	
Automotive – service/repair	0.70
Bank/office – banking activity area	1.5
Dormitory living quarters	1.10
Gymnasium/fitness center	
Fitness area	0.9
Gymnasium audience/seating	0.40
Playing area	1.40

(continued)

TABLE C405.5.2(2)—continued
INTERIOR LIGHTING POWER ALLOWANCES:
SPACE-BY-SPACE METHOD

COMMON SPACE-BY-SPACE TYPES	LPD (w/ft²)
Healthcare clinic/hospital	
Corridors/transition	1.00
Exam/treatment	1.70
Emergency	2.70
Public and staff lounge	0.80
Medical supplies	1.40
Nursery	0.9
Nurse station	1.00
Physical therapy	0.90
Patient room	0.70
Pharmacy	1.20
Radiology/imaging	1.3
Operating room	2.20
Recovery	1.2
Lounge/recreation	0.8
Laundry – washing	0.60
Hotel	
Dining area	1.30
Guest rooms	1.10
Hotel lobby	2.10
Highway lodging dining	1.20
Highway lodging guest rooms	1.10
Library	
Stacks	1.70
Card file and cataloguing	1.10
Reading area	1.20
Manufacturing	
Corridors/transition	0.40
Detailed manufacturing	1.3
Equipment room	1.0
Extra high bay (> 50-foot floor-ceiling height)	1.1
High bay (25- – 50-foot floor-ceiling height)	1.20
Low bay (< 25-foot floor-ceiling height)	1.2
Museum	
General exhibition	1.00
Restoration	1.70
Parking garage – garage areas	0.2
Convention center	
Exhibit space	1.50
Audience/seating area	0.90
Fire stations	
Engine room	0.80
Sleeping quarters	0.30
Post office	
Sorting area	0.9
Religious building	
Fellowship hall	0.60
Audience seating	2.40
Worship pulpit/choir	2.40
Retail	
Dressing/fitting area	0.9
Mall concourse	1.6
Sales area	1.6[a]

(continued)

TABLE C405.5.2(2)—continued
INTERIOR LIGHTING POWER ALLOWANCES:
SPACE-BY-SPACE METHOD

BUILDING SPECIFIC SPACE-BY-SPACE TYPES	LPD (w/ft²)
Sports arena	
Audience seating	0.4
Court sports area – Class 4	0.7
Court sports area – Class 3	1.2
Court sports area – Class 2	1.9
Court sports area – Class 1	3.0
Ring sports area	2.7
Transportation	
Air/train/bus baggage area	1.00
Airport concourse	0.60
Terminal – ticket counter	1.50
Warehouse	
Fine material storage	1.40
Medium/bulky material	0.60

For SI: 1 foot = 304.8 mm, 1 watt per square foot = 11 W/m².

a. Where lighting equipment is specified to be installed to highlight specific merchandise in addition to lighting equipment specified for general lighting and is switched or dimmed on circuits different from the circuits for general lighting, the smaller of the actual wattage of the lighting equipment installed specifically for merchandise, or additional lighting power as determined below shall be added to the interior lighting power determined in accordance with this line item.

Calculate the additional lighting power as follows:

Additional Interior Lighting Power Allowance = 500 watts + (Retail Area 1 × 0.6 W/ft²) + (Retail Area 2 × 0.6 W/ft²) + (Retail Area 3 × 1.4 W/ft²) + (Retail Area 4 × 2.5 W/ft²).

where:

Retail Area 1 = The floor area for all products not listed in Retail Area 2, 3 or 4.

Retail Area 2 = The floor area used for the sale of vehicles, sporting goods and small electronics.

Retail Area 3 = The floor area used for the sale of furniture, clothing, cosmetics and artwork.

Retail Area 4 = The floor area used for the sale of jewelry, crystal and china.

Exception: Other merchandise categories are permitted to be included in Retail Areas 2 through 4 above, provided that justification documenting the need for additional lighting power based on visual inspection, contrast, or other critical display is *approved* by the authority having jurisdiction.

C405.6.2 Exterior building lighting power. The total exterior lighting power allowance for all exterior building applications is the sum of the base site allowance plus the individual allowances for areas that are to be illuminated and are permitted in Table C405.6.2(2) for the applicable lighting zone. Tradeoffs are allowed only among exterior lighting applications listed in Table C405.6.2(2), Tradable Surfaces section. The lighting zone for the building exterior is determined from Table C405.6.2(1) unless otherwise specified by the local jurisdiction. Exterior lighting for all applications (except those included in the exceptions to Section C405.6.2) shall comply with the requirements of Section C405.6.1.

Exception: Lighting used for the following exterior applications is exempt where equipped with a control device independent of the control of the nonexempt lighting:

1. Specialized signal, directional and marker lighting associated with transportation;
2. Advertising signage or directional signage;
3. Integral to equipment or instrumentation and is installed by its manufacturer;
4. Theatrical purposes, including performance, stage, film production and video production;
5. Athletic playing areas;
6. Temporary lighting;
7. Industrial production, material handling, transportation sites and associated storage areas;
8. Theme elements in theme/amusement parks; and
9. Used to highlight features of public monuments and registered historic landmark structures or buildings.

TABLE C405.6.2(1)
EXTERIOR LIGHTING ZONES

LIGHTING ZONE	DESCRIPTION
1	Developed areas of national parks, state parks, forest land, and rural areas
2	Areas predominantly consisting of residential zoning, neighborhood business districts, light industrial with limited nighttime use and residential mixed use areas
3	All other areas
4	High-activity commercial districts in major metropolitan areas as designated by the local land use planning authority

C405.7 Electrical energy consumption (Mandatory). In buildings having individual dwelling units, provisions shall be made to determine the electrical energy consumed by each tenant by separately metering individual dwelling units.

SECTION C406
ADDITIONAL EFFICIENCY PACKAGE OPTIONS

C406.1 Requirements. Buildings shall comply with at least one of the following:

1. Efficient HVAC Performance in accordance with Section C406.2.
2. Efficient Lighting System in accordance with Section C406.3.
3. On-Site Supply of Renewable Energy in accordance with Section C406.4.

Individual tenant spaces shall comply with either Section C406.2 or Section C406.3 unless documentation can be provided that demonstrates compliance with Section C406.4 for the entire building

TABLE C405.6.2(2)
INDIVIDUAL LIGHTING POWER ALLOWANCES FOR BUILDING EXTERIORS

		LIGHTING ZONES			
		Zone 1	Zone 2	Zone 3	Zone 4
Base Site Allowance (Base allowance is usable in tradable or nontradable surfaces.)		500 W	600 W	750 W	1300 W
Tradable Surfaces (Lighting power densities for uncovered parking areas, building grounds, building entrances and exits, canopies and overhangs and outdoor sales areas are tradable.)	**Uncovered Parking Areas**				
	Parking areas and drives	0.04 W/ft^2	0.06 W/ft^2	0.10 W/ft^2	0.13 W/ft^2
	Building Grounds				
	Walkways less than 10 feet wide	0.7 W/linear foot	0.7 W/linear foot	0.8 W/linear foot	1.0 W/linear foot
	Walkways 10 feet wide or greater, plaza areas special feature areas	0.14 W/ft^2	0.14 W/ft^2	0.16 W/ft^2	0.2 W/ft^2
	Stairways	0.75 W/ft^2	1.0 W/ft^2	1.0 W/ft^2	1.0 W/ft^2
	Pedestrian tunnels	0.15 W/ft^2	0.15 W/ft^2	0.2 W/ft^2	0.3 W/ft^2
	Building Entrances and Exits				
	Main entries	20 W/linear foot of door width	20 W/linear foot of door width	30 W/linear foot of door width	30 W/linear foot of door width
	Other doors	20 W/linear foot of door width	20 W/linear foot of door width	20 W/linear foot of door width	20 W/linear foot of door width
	Entry canopies	0.25 W/ft^2	0.25 W/ft^2	0.4 W/ft^2	0.4 W/ft^2
	Sales Canopies				
	Free-standing and attached	0.6 W/ft^2	0.6 W/ft^2	0.8 W/ft^2	1.0 W/ft^2
	Outdoor Sales				
	Open areas (including vehicle sales lots)	0.25 W/ft^2	0.25 W/ft^2	0.5 W/ft^2	0.7 W/ft^2
	Street frontage for vehicle sales lots in addition to "open area" allowance	No allowance	10 W/linear foot	10 W/linear foot	30 W/linear foot
Nontradable Surfaces (Lighting power density calculations for the following applications can be used only for the specific application and cannot be traded between surfaces or with other exterior lighting. The following allowances are in addition to any allowance otherwise permitted in the "Tradable Surfaces" section of this table.)	Building facades	No allowance	0.1 W/ft^2 for each illuminated wall or surface or 2.5 W/linear foot for each illuminated wall or surface length	0.15 W/ft^2 for each illuminated wall or surface or 3.75 W/linear foot for each illuminated wall or surface length	0.2 W/ft^2 for each illuminated wall or surface or 5.0 W/linear foot for each illuminated wall or surface length
	Automated teller machines and night depositories	270 W per location plus 90 W per additional ATM per location	270 W per location plus 90 W per additional ATM per location	270 W per location plus 90 W per additional ATM per location	270 W per location plus 90 W per additional ATM per location
	Entrances and gatehouse inspection stations at guarded facilities	0.75 W/ft^2 of covered and uncovered area	0.75 W/ft^2 of covered and uncovered area	0.75 W/ft^2 of covered and uncovered area	0.75 W/ft^2 of covered and uncovered area
	Loading areas for law enforcement, fire, ambulance and other emergency service vehicles	0.5 W/ft^2 of covered and uncovered area	0.5 W/ft^2 of covered and uncovered area	0.5 W/ft^2 of covered and uncovered area	0.5 W/ft^2 of covered and uncovered area
	Drive-up windows/doors	400 W per drive-through	400 W per drive-through	400 W per drive-through	400 W per drive-through
	Parking near 24-hour retail entrances	800 W per main entry	800 W per main entry	800 W per main entry	800 W per main entry

For SI: 1 foot = 304.8 mm, 1 watt per square foot = W/0.0929 m^2.

TABLE C406.2(1)
UNITARY AIR CONDITIONERS AND CONDENSING UNITS, ELECTRICALLY OPERATED, EFFICIENCY REQUIREMENTS

EQUIPMENT TYPE	SIZE CATEGORY	SUBCATEGORY OR RATING CONDITION	MINIMUM EFFICIENCY[a]	
			CLIMATE ZONES 1 - 5	CLIMATE ZONES 6 - 8
Air conditioners, air cooled	< 65,000 Btu/h	Split system	15.0 SEER 12.5 EER	14 SEER 12 EER
		Single package	15.0 SEER 12.0 EER	14.0 SEER 11.6 EER
	≥ 65,000 Btuh/h and < 240,000 Btu/h	Split system and single package	12.0 EER[b] 12.54 IEER[b]	11.5 EER[b] 12.0 IEER[b]
	≥ 240,000 Btu/h and < 760,000 Btu/h	Split system and single package	10.8 EER[b] 11.3 IEER[b]	10.5 EER[b] 11.0 IEER[b]
	≥ 760,000 Btu/h	—	10.2 EER[b] 10.7 IEER[b]	9.7 EER[b] 10.2 IEER[b]
Air conditioners, water and evaporatively cooled	—	Split system and single package	14.0 EER	14.0 EER

For SI: 1 British thermal unit per hour = 0.2931 W.

a. IEERs are only applicable to equipment with capacity modulation.

b. Deduct 0.2 from the required EERs and IPLVs for units with a heating section other than electric resistance heat.

TABLE C406.2(2)
UNITARY AND APPLIED HEAT PUMPS, ELECTRICALLY OPERATED, EFFICIENCY REQUIREMENTS

EQUIPMENT TYPE	SIZE CATEGORY	SUBCATEGORY OR RATING CONDITION	MINIMUM EFFICIENCY[a]	
			CLIMATE ZONES 1 - 5	CLIMATE ZONES 6 - 8
Air cooled (Cooling mode)	< 65,000 Btu/h	Split system	15.0 SEER, 12.5 EER	14.0 SEER, 12.0 EER
		Single package	15.0 SEER, 12.0 EER	14.0 SEER 11.6 EER
	≥ 65,000 Btu/h and < 240,000 Btu/h	Split system and single package	12.0 SEER, 12.4 EER	11.5 EER[b], 12.0 IEER[b]
	≥ 240,000 Btu/h	Split system and single package	12.0 SEER, 12.4 EER	10.5 EER[b], 10.5 IEER[b]
Water sources (Cooling mode)	< 135,000 Btu/h	85°F entering water	14.0 EER	14.0 EER
Air cooled (Heating mode)	< 65,000 Btu/h (Cooling capacity)	Split system	9.0 HSPF	8.5 HSPF
		Single package	8.5 HSPF	8.0 HSPF
	≥ 65,000 Btu/h and < 135,000 Btu/h (Cooling capacity)	47°F db/43°F wb outdoor air	3.4 COP	3.4 COP
		17°F db/15°F wb outdoor air	2.4 COP	2.4 COP
	≥ 135,000 Btu/h (Cooling capacity)	47°F db/43°F wb outdoor air	3.2 COP	3.2 COP
		77°F db/15°F wb outdoor air	2.1 COP	2.1 COP
Water sources (Heating mode)	< 135,000 Btu/h (Cooling capacity)	70°F entering water	4.6 COP	4.6 COP

For SI: °C = [(°F) - 32] / 1.8, 1 British thermal unit per hour = 0.2931 W.

db = dry-bulb temperature, °F; wb = wet-bulb temperature, °F.

a. IEERs and Part load rating conditions are only applicable to equipment with capacity modulation.

b. Deduct 0.2 from the required EERs and IPLVs for units with a heating section other than electric resistance heat.

TABLE C406.2(3)
PACKAGED TERMINAL AIR CONDITIONERS AND PACKAGED TERMINALHEAT PUMPS

EQUIPMENT TYPE	SIZE CATEGORY	MINIMUM EFFICIENCY
Air conditioners and heat pumps (cooling mode)	< 7,000 Btu/h	11.9 EER
	7,000 Btu/h and < 10,000 Btu/h	11.3 EER
	10,000 Btu/h and ≤ 13,000 Btu/h	10.7 EER
	> 13,000 Btu/h	9.5 EER

TABLE C406.2(4)
WARM AIR FURNACES AND COMBINATION WARM AIR FURNACES/AIR-CONDITIONING UNITS,
WARM AIR DUCT FURNACES AND UNIT HEATERS, EFFICIENCY REQUIREMENTS

EQUIPMENT TYPE	SIZE CATEGORY (INPUT)	SUBCATEGORY OR RATING CONDITION	MINIMUM EFFICIENCY	TEST PROCEDURE
Warm air furnaces, gas fired[a]	< 225,000 Btu/h	—	For Climate Zones 1 and 2 NR	DOE 10 CFR Part 430 or ANSI Z21.47
			For Climate Zones 3 and 4 90 AFUE or 90 E_t^c	
			For Climate Zones 4 – 8 92 AFUE or 92 E_t^c	
	≥ 225,000 Btu/h	Maximum capacity	90% E_c^b	ANSI Z21.47
Warm air furnaces, oil fired[a]	< 225,000 Btu/h	—	For Climate Zones 1 and 2 NR	DOE 10 CFR Part 430 or UL 727
			For Climate Zones 3 – 8 85 AFUE or 85 E_t^c	
	≥ 225,000 Btu/h	Maximum capacity	85% E_t^b	UL 727
Warm air duct furnaces, gas fired[a]	All capacities	Maximum capacity	90% E_c	ANSI Z83.8
Warm air unit heaters, gas fired	All capacities	Maximum capacity	90% E_c	ANSI Z83.8
Warm air unit heaters, oil fired	All capacities	Maximum capacity	90% E_c	UL 731

For SI: 1 British thermal unit per hour = 0.2931 W.

E_t = Thermal efficiency. E_c = Combustion efficiency (100 percent less flue losses).

a. Efficient furnace fan: Fossil fuel furnaces in climate zones 3 to 8 shall have a furnace electricity ratio not greater than 2 percent and shall include a manufacturer's designation of the furnace electricity ratio.

b. Units shall also include an IID (intermittent ignition device), have jacket losses not exceeding 0.75 percent of the input rating, and have either power venting or a flue damper. A vent damper is an acceptable alternative to a flue damper for those furnaces where combustion air is drawn from the conditioned space.

c. Where there are two ratings for units not covered by NAECA (3-phase power or cooling capacity greater than or equal to 65,000 Btu/h [19 kW]), units shall be permitted to comply with either rating.

TABLE C406.2(5)
BOILER, EFFICIENCY REQUIREMENTS

EQUIPMENT TYPE	FUEL	SIZE CATEGORY	TEST PROCEEDURE	MINIMUM EFFICIENCY
Steam	Gas	< 300,000 Btu/h	DOE 10 CFR Part 430	83% AFUE
		> 300,000 Btu / h and > 2.5 m Btu/h	DOE 10 CFR Part 431	81% E_t
		>2.5 m Btu/h		82% E_c
	Oil	< 300,000 Btu/h	DOE 10 CFR Part 430	85% AFUE
		> 300,000 Btu/h and > 2.5 m Btu/h	DOE 10 CFR Part 431	83% E_t
		>2.5 m Btu/h		84% E_c
Hot water	Gas	< 300,000 Btu/h	DOE 10 CFR Part 430	97% AFUE
		> 300,000 Btu/h and > 2.5 m Btu/h	DOE 10 CFR Part 431	97% E_t
		>2.5 m Btu/h		94% E_c
	Oil	< 300,000 Btu/h	DOE 10 CFR Part 430	90% AFUE
		> 300,000 Btu/h and > 2.5 m Btu/h	DOE 10 CFR Part 431	88% E_t
		>2.5 m Btu/h		87% E_c

For SI: 1 British thermal unit per hour = 0.2931 W.

E_t = Thermal efficiency. E_c = Combustion efficiency (100 percent less flue losses).

TABLE C406.2(6)
CHILLERS—EFFICIENCY REQUIREMENTS

EQUIPMENT TYPE	SIZE CATEGORY	UNITS	MINIMUM EFFICIENCY[c] (I-P)				Test Procedure[b]
			Path A		Path B[c]		
			Full Load	IPLV	Full Load	IPLV	
Air-cooled chillers with condenser, electrically operated	< 150 tons	EER	10.000	12.500	NA	NA	AHRI 550/590[f]
	≥ 150 tons	EER	10.000	12.750	NA	NA	
Air-cooled without condenser, electrical operated	All capacities	EER	Condenserless units shall be rated with matched condensers				AHRI 550/590[f]
Water-cooled, electrically operated, positive displacement (reciprocating)	All capacities	kw/ton	Reciprocating units required to comply with water cooled positive displacement requirements				AHRI 550/590[f]
Water-cooled electrically operated, positive displacement	< 75 tons	kw/ton	0.780	0.630	0.800	0.600	AHRI 550/590[f]
	≥ 75 tons and < 150 tons	kw/ton	0.775	0.615	0.790	0.586	
	≥ 150 tons and < 300 tons	kw/ton	0.680	0.580	0.718	0.540	
	≥ 300 tons	kw/ton	0.620	0.540	0.639	0.490	
Water-cooled electrically operated, centrifugal[d]	< 150 tons	kw/ton	0.634	0.596	0.639	0.450	AHRI 550/590[f]
	≥ 150 tons and < 300 tons	kw/ton	0.634	0.596	0.639	0.450	
	≥ 300 tons and < 600 tons	kw/ton	0.576	0.549	0.600	0.400	
	≥ 600 tons	kw/ton	0.570	0.539	0.590	0.400	
Air-cooled absorption single effect[e]	All capacities	COP	0.600	NR	NA	NA	AHRI 560
Water-cooled absorption single effect[e]	All capacities	COP	0.700	NR	NA	NA	
Absorption double effect indirect-fired	All capacities	COP	1.000	1.050	NA	NA	
Absorption double effect direct fired	All capacities	COP	1.000	1.000	NA	NA	

For SI: 1 Ton = 3516 W.

NA = Not applicable and cannot be used for compliance.   NR = No minimum requirements.

a. Compliance with this standard can be obtained by meeting the minimum requirements of Path A or Path B. However both the full load and IPLV shall be met to fulfill the requirements of Path A and Path B.

b. Chapter 6 of the referenced standard contains a complete specification of the referenced test procedure, including the referenced year version of the test procedure.

c. Path B is intended for applications with significant operating time at part load. All Path B machines shall be equipped with demand limiting capable controls.

d. The chiller equipment requirements do not apply for chillers used in low-temperature applications where the design leaving fluid temperature is greater than 40°F.

e. Only allowed to be used in heat recovery applications.

f. Packages that are not designed for operation at ARI Standard 550/590 test conditions (and, thus, cannot be tested to meet the requirements of Table C-3) of 44°F leaving chilled-water temperature and 85°F entering condenser-water temperature with 3 gpm/ton condenser-water flow shall have maximum full-load kW/ton and *NPLV* ratings adjusted using the following equation:

Adjusted maximum full load kW/ton rating = (full load kW/ton from Table C-3)/K_{adj}

Adjusted maximum NPLV rating = (IPLV from Table C-3)/K_{adj}

where:

K_{adj} = $6.174722 - 0.303668(X) + 0.00629466(X)^2 - 0.000045780(X)^3$

X = DT_{std} + LIFT (°F)

DT_{std} = [(24 + (full load kW/ton from Table C-3) × 6.83)]/flow (°F)

Flow = condenser-water flow (gpm) / cooling full load capacity (tons)

LIFT = CEWT – CLWT (°F)

CEWT = full load entering condenser-water temperature (°F)

CLWT = full load leaving chilled-water temperature (°F)

The adjusted full load and *NPLV* values are only applicable over the following full-load design ranges:

Minimum leaving chilled-water temperature: 38°F

Maximum condenser entering water temperature: 102°F

Condenser-water flow: 1 to 6 gpm/ton

$X \geq 39°F$ and $\leq 60°F$

TABLE C406.2(7)
ABSORPTION CHILLERS—EFFICIENCY REQUIREMENTS

EQUIPMENT TYPE	MINIMUM EFFICIENCY FULL LOAD COP (IPLV)
Air cooled, single effect	0.60, allowed only in heat recovery applications
Water cooled, single effect	0.70, allowed only in heat recovery applications
Double effect – direct fired	1.0 (1.05)
Double effect – indirect fired	1.20

C406.2 Efficient HVAC performance. Equipment shall meet the minimum efficiency requirements of Tables C406.2.(1) through C406.2(7) in addition to the requirements in Section C403. This section shall only be used where the equipment efficiencies in Tables C406.2(1) through C406.2(7) are greater than the equipment efficiencies listed in Table C403.2.3(1) through 403.2.3(7) for the equipment type.

C406.3 Efficient lighting system. Whole building lighting power density (Watts/sf) shall comply with the requirements of Section C406.3.1.

C406.3.1 Reduced lighting power density. The total interior lighting power (watts) of the building shall be determined by using the reduced whole building interior lighting power in Table C406.3 times the floor area for the building types.

C406.4 On-site renewable energy. Total minimum ratings of on-site renewable energy systems shall comply with one of the following:

1. Provide not less than 1.75 Btu (1850 W), or not less than 0.50 watts per square foot (5.4 W/m^2) of conditioned floor area.

2. Provide not less than 3 percent of the energy used within the building for building mechanical and service water heating equipment and lighting regulated in this chapter.

SECTION C407
TOTAL BUILDING PERFORMANCE

C407.1 Scope. This section establishes criteria for compliance using total building performance. The following systems and loads shall be included in determining the total building performance: heating systems, cooling systems, service water heating, fan systems, lighting power, receptacle loads and process loads.

C407.2 Mandatory requirements. Compliance with this section requires that the criteria of Sections C402.4, C403.2, C404 and C405 be met.

C407.3 Performance-based compliance. Compliance based on total building performance requires that a proposed building (*proposed design*) be shown to have an annual energy cost that is less than or equal to the annual energy cost of the *standard reference design*. Energy prices shall be taken from a source *approved* by the *code official,* such as the Department of Energy, Energy Information Administration's *State*

Energy Price and Expenditure Report. Code officials shall be permitted to require time-of-use pricing in energy cost calculations. Nondepletable energy collected off site shall be treated and priced the same as purchased energy. Energy from nondepletable energy sources collected on site shall be omitted from the annual energy cost of the *proposed design.*

Exception: Jurisdictions that require site energy (1 kWh = 3413 Btu) rather than energy cost as the metric of comparison.

TABLE C406.3
REDUCED INTERIOR LIGHTING POWER

BUILDING AREA TYPE[a]	LPD (w/ft^2)
Automotive facility	0.82
Convention center	1.08
Courthouse	1.05
Dining: bar lounge/leisure	0.99
Dining: cafeteria/fast food	0.90
Dining: family	0.89
Dormitory	0.61
Exercise center	0.88
Fire station	0.71
Gymnasium	1.0
Health care clinic	0.87
Hospital	1.10
Library	1.18
Manufacturing facility	1.11
Hotel/motel	0.88
Motion picture theater	0.83
Museum	1.06
Multifamily	0.60
Office	0.90/0.85[b]
Performing arts theater	1.39
Police station	0.96
Post office	0.87
Religious building	1.05
Retail	1.4/1.3[b]
School/ university	0.99
Sports arena	0.78
Town hall	0.92
Transportation	0.77
Warehouse[c]	0.6
Workshop	1.2

For SI: 1 foot = 304.8 mm, 1 watt per square foot = W/0.0929 m^2.

a. In cases where both a general building area type and a more specific building area type are listed, the more specific building area type shall apply.

b. First LPD value applies if no less than 30 percent of conditioned floor area is in daylight zones. Automatic daylighting controls shall be installed in daylight zones and shall meet the requirements of Section C405.2.2.3. In all other cases, second LPD value applies.

c. No less than 70 percent of the floor area shall be in the daylight zone. Automatic daylighting controls shall be installed in daylight zones and shall meet the requirements of Section C405.2.2.3.

C407.4 Documentation. Documentation verifying that the methods and accuracy of compliance software tools conform to the provisions of this section shall be provided to the *code official.*

C407.4.1 Compliance report. Compliance software tools shall generate a report that documents that the *proposed design* has annual energy costs less than or equal to the annual energy costs of the *standard reference design.* The compliance documentation shall include the following information:

1. Address of the building;

2. An inspection checklist documenting the building component characteristics of the *proposed design* as *listed* in Table C407.5.1(1). The inspection checklist shall show the estimated annual energy cost for both the *standard reference design* and the *proposed design;*

3. Name of individual completing the compliance report; and

4. Name and version of the compliance software tool.

C407.4.2 Additional documentation. The *code official* shall be permitted to require the following documents:

1. Documentation of the building component characteristics of the *standard reference design;*

2. Thermal zoning diagrams consisting of floor plans showing the thermal zoning scheme for *standard reference design* and *proposed design;*

3. Input and output report(s) from the energy analysis simulation program containing the complete input and output files, as applicable. The output file shall include energy use totals and energy use by energy source and end-use served, total hours that space conditioning loads are not met and any errors or warning messages generated by the simulation tool as applicable;

4. An explanation of any error or warning messages appearing in the simulation tool output; and

5. A certification signed by the builder providing the building component characteristics of the *proposed design* as given in Table C407.5.1(1).

C407.5 Calculation procedure. Except as specified by this section, the *standard reference design* and *proposed design* shall be configured and analyzed using identical methods and techniques.

C407.5.1 Building specifications. The *standard reference design* and *proposed design* shall be configured and analyzed as specified by Table C407.5.1(1). Table C407.5.1(1) shall include by reference all notes contained in Table C402.2.

C407.5.2 Thermal blocks. The *standard reference design* and *proposed design* shall be analyzed using identical ther-

mal blocks as specified in Section C407.5.2.1, C407.5.2.2 or C407.5.2.3.

C407.5.2.1 HVAC zones designed. Where HVAC *zones* are defined on HVAC design drawings, each HVAC *zone* shall be modeled as a separate thermal block.

Exception: Different HVAC *zones* shall be allowed to be combined to create a single thermal block or identical thermal blocks to which multipliers are applied provided:

1. The space use classification is the same throughout the thermal block.

2. All HVAC *zones* in the thermal block that are adjacent to glazed exterior walls face the same orientation or their orientations are within 45 degrees (0.79 rad) of each other.

3. All of the *zones* are served by the same HVAC system or by the same kind of HVAC system.

C407.5.2.2 HVAC zones not designed. Where HVAC *zones* have not yet been designed, thermal blocks shall be defined based on similar internal load densities, occupancy, lighting, thermal and temperature schedules, and in combination with the following guidelines:

1. Separate thermal blocks shall be assumed for interior and perimeter spaces. Interior spaces shall be those located more than 15 feet (4572 mm) from an exterior wall. Perimeter spaces shall be those located closer than 15 feet (4572 mm) from an *exterior wall.*

2. Separate thermal blocks shall be assumed for spaces adjacent to glazed exterior walls: a separate *zone* shall be provided for each orientation, except orientations that differ by no more than 45 degrees (0.79 rad) shall be permitted to be considered to be the same orientation. Each *zone* shall include floor area that is 15 feet (4572 mm) or less from a glazed perimeter wall, except that floor area within 15 feet (4572 mm) of glazed perimeter walls having more than one orientation shall be divided proportionately between *zones.*

3. Separate thermal blocks shall be assumed for spaces having floors that are in contact with the ground or exposed to ambient conditions from *zones* that do not share these features.

4. Separate thermal blocks shall be assumed for spaces having exterior ceiling or roof assemblies from *zones* that do not share these features.

C407.5.2.3 Multifamily residential buildings. Residential spaces shall be modeled using one thermal block per space except that those facing the same orientations are permitted to be combined into one thermal block. Corner units and units with roof or floor loads shall only be combined with units sharing these features.

TABLE C407.5.1(1)
SPECIFICATIONS FOR THE STANDARD REFERENCE AND PROPOSED DESIGNS

BUILDING COMPONENT CHARACTERISTICS	STANDARD REFERENCE DESIGN	PROPOSED DESIGN
Space use classification	Same as proposed	The space use classification shall be chosen in accordance with Table C405.5.2 for all areas of the building covered by this permit. Where the space use classification for a building is not known, the building shall be categorized as an office building.
Roofs	Type: Insulation entirely above deck	As proposed
	Gross area: same as proposed	As proposed
	U-factor: from Table C402.1.2	As proposed
	Solar absorptance: 0.75	As proposed
	Emittance: 0.90	As proposed
Walls, above-grade	Type: Mass wall if proposed wall is mass; otherwise steel-framed wall	As proposed
	Gross area: same as proposed	As proposed
	U-factor: from Table C402.1.2	As proposed
	Solar absorptance: 0.75	As proposed
	Emittance: 0.90	As proposed
Walls, below-grade	Type: Mass wall	As proposed
	Gross area: same as proposed	As proposed
	U-Factor: from Table C402.1.2 with insulation layer on interior side of walls	As proposed
Floors, above-grade	Type: joist/framed floor	As proposed
	Gross area: same as proposed	As proposed
	U-factor: from Table C402.1.2	As proposed
Floors, slab-on-grade	Type: Unheated	As proposed
	F-factor: from Table C402.1.2	As proposed
Doors	Type: Swinging	As proposed
	Area: Same as proposed	As proposed
	U-factor: from Table C402.2	As proposed
Glazing	Area 1. The proposed glazing area; where the proposed glazing area is less than 40 percent of above-grade wall area. 2. 40 percent of above-grade wall area; where the proposed glazing area is 40 percent or more of the above-grade wall area.	As proposed
	U-factor: from Table C402.3	As proposed
	SHGC: from Table C402.3 except that for climates with no requirement (NR) SHGC = 0.40 shall be used	As proposed
	External shading and PF: None	As proposed
Skylights	Area 1. The proposed skylight area; where the proposed skylight area is less than 3 percent of gross area of roof assembly. 2. 3 percent of gross area of roof assembly; where the proposed skylight area is 3 percent or more of gross area of roof assembly	As proposed
	U-factor: from Table C402.3	As proposed
	SHGC: from Table C402.3 except that for climates with no requirement (NR) SHGC = 0.40 shall be used.	As proposed
Lighting, interior	The interior lighting power shall be determined in accordance with Section C405.5.2. Where the occupancy of the building is not known, the lighting power density shall be 1.0 Watt per square foot (10.73 W/m^2) based on the categorization of buildings with unknown space classification as offices.	As proposed
Lighting, exterior	The lighting power shall be determined in accordance with Table C405.6.2(2). Areas and dimensions of tradable and nontradable surfaces shall be the same as proposed.	As proposed

(continued)

TABLE C407.5.1(1)—continued
SPECIFICATIONS FOR THE STANDARD REFERENCE AND PROPOSED DESIGNS

BUILDING COMPONENT CHARACTERISTICS	STANDARD REFERENCE DESIGN	PROPOSED DESIGN
Internal gains	Same as proposed	Receptacle, motor and process loads shall be modeled and estimated based on the space use classification. All end-use load components within and associated with the building shall be modeled to include, but not be limited to, the following: exhaust fans, parking garage ventilation fans, exterior building lighting, swimming pool heaters and pumps, elevators, escalators, refrigeration equipment and cooking equipment.
Schedules	Same as proposed	Operating schedules shall include hourly profiles for daily operation and shall account for variations between weekdays, weekends, holidays and any seasonal operation. Schedules shall model the time-dependent variations in occupancy, illumination, receptacle loads, thermostat settings, mechanical ventilation, HVAC equipment availability, service hot water usage and any process loads. The schedules shall be typical of the proposed building type as determined by the designer and approved by the jurisdiction.
Mechanical ventilation	Same as proposed	As proposed, in accordance with Section C403.2.5.
Heating systems	Fuel type: same as proposed design	As proposed
	Equipment type[a]: from Tables C407.5.1(2) and C407.5.1(3)	As proposed
	Efficiency: from Tables C403.2.3(4) and C403.2.3(5)	As proposed
	Capacity[b]: sized proportionally to the capacities in the proposed design based on sizing runs, and shall be established such that no smaller number of unmet heating load hours and no larger heating capacity safety factors are provided than in the proposed design.	As proposed
Cooling systems	Fuel type: same as proposed design	As proposed
	Equipment type[c]: from Tables C407.5.1(2) and C407.5.1(3)	As proposed
	Efficiency: from Tables C403.2.3(1), C403.2.3(2) and C403.2.3(3)	As proposed
	Capacity[b]: sized proportionally to the capacities in the proposed design based on sizing runs, and shall be established such that no smaller number of unmet cooling load hours and no larger cooling capacity safety factors are provided than in the proposed design.	As proposed
	Economizer[d]: same as proposed, in accordance with Section C403.4.1.	As proposed
Service water heating	Fuel type: same as proposed	As proposed
	Efficiency: from Table C404.2	As proposed
	Capacity: same as proposed	As proposed
	Where no service water hot water system exists or is specified in the proposed design, no service hot water heating shall be modeled.	

a. Where no heating system exists or has been specified, the heating system shall be modeled as fossil fuel. The system characteristics shall be identical in both the standard reference design and proposed design.

b. The ratio between the capacities used in the annual simulations and the capacities determined by sizing runs shall be the same for both the standard reference design and proposed design.

c. Where no cooling system exists or no cooling system has been specified, the cooling system shall be modeled as an air-cooled single-zone system, one unit per thermal zone. The system characteristics shall be identical in both the standard reference design and proposed design.

d. If an economizer is required in accordance with Table C403.3.1(1), and if no economizer exists or is specified in the proposed design, then a supply air economizer shall be provided in accordance with Section C403.4.1.

TABLE C407.5.1(2)
HVAC SYSTEMS MAP

CONDENSER COOLING SOURCE[a]	HEATING SYSTEM CLASSIFICATION[b]	STANDARD REFERENCE DESIGN HVC SYSTEM TYPE[c]		
		Single-zone Residential System	Single-zone Nonresidential System	All Other
Water/ground	Electric resistance	System 5	System 5	System 1
	Heat pump	System 6	System 6	System 6
	Fossil fuel	System 7	System 7	System 2
Air/none	Electric resistance	System 8	System 9	System 3
	Heat pump	System 8	System 9	System 3
	Fossil fuel	System 10	System 11	System 4

a. Select "water/ground" if the proposed design system condenser is water or evaporatively cooled; select "air/none" if the condenser is air cooled. Closed-circuit dry coolers shall be considered air cooled. Systems utilizing district cooling shall be treated as if the condenser water type were "water." If no mechanical cooling is specified or the mechanical cooling system in the proposed design does not require heat rejection, the system shall be treated as if the condenser water type were "Air." For proposed designs with ground-source or groundwater-source heat pumps, the standard reference design HVAC system shall be water-source heat pump (System 6).

b. Select the path that corresponds to the proposed design heat source: electric resistance, heat pump (including air source and water source), or fuel fired. Systems utilizing district heating (steam or hot water) and systems with no heating capability shall be treated as if the heating system type were "fossil fuel." For systems with mixed fuel heating sources, the system or systems that use the secondary heating source type (the one with the smallest total installed output capacity for the spaces served by the system) shall be modeled identically in the standard reference design and the primary heating source type shall be used to determine *standard* reference design HVAC system type.

c. Select the standard reference design HVAC system category: The system under "single-zone residential system" shall be selected if the HVAC system in the proposed design is a single-zone system and serves a residential space. The system under "single-zone nonresidential system" shall be selected if the HVAC system in the proposed design is a single-zone system and serves other than residential spaces. The system under "all other" shall be selected for all other cases.

TABLE C407.5.1(3)
SPECIFICATIONS FOR THE STANDARD REFERENCE DESIGN HVAC SYSTEM DESCRIPTIONS

SYSTEM NO.	SYSTEM TYPE	FAN CONTROL	COOLING TYPE	HEATING TYPE
1	Variable air volume with parallel fan-powered boxes[a]	VAV[d]	Chilled water[e]	Electric resistance
2	Variable air volume with reheat[b]	VAV[d]	Chilled water[e]	Hot water fossil fuel boiler[f]
3	Packaged variable air volume with parallel fan-powered boxes[a]	VAV[d]	Direct expansion[c]	Electric resistance
4	Packaged variable air volume with reheat[b]	VAV[d]	Direct expansion[c]	Hot water fossil fuel boiler[f]
5	Two-pipe fan coil	Constant volume[i]	Chilled water[e]	Electric resistance
6	Water-source heat pump	Constant volume[i]	Direct expansion[c]	Electric heat pump and boiler[g]
7	Four-pipe fan coil	Constant volume[i]	Chilled water[e]	Hot water fossil fuel boiler[f]
8	Packaged terminal heat pump	Constant volume[i]	Direct expansion[c]	Electric heat pump[h]
9	Packaged rooftop heat pump	Constant volume[i]	Direct expansion[c]	Electric heat pump[h]
10	Packaged terminal air conditioner	Constant volume[i]	Direct expansion	Hot water fossil fuel boiler[f]
11	Packaged rooftop air conditioner	Constant volume[i]	Direct expansion	Fossil fuel furnace

For SI: 1 foot = 304.8 mm, 1 cfm/ft^2 = 0.0004719, 1 Btu/h = 0.293/W, °C = [(°F) -32/1.8].

a. **VAV with parallel boxes:** Fans in parallel VAV fan-powered boxes shall be sized for 50 percent of the peak design flow rate and shall be modeled with 0.35 W/cfm fan power. Minimum volume setpoints for fan-powered boxes shall be equal to the minimum rate for the space required for ventilation consistent with Section C403.4.5, Exception 5. Supply air temperature setpoint shall be constant at the design condition.

b. **VAV with reheat:** Minimum volume setpoints for VAV reheat boxes shall be 0.4 cfm/ft^2 of floor area. Supply air temperature shall be reset based on zone demand from the design temperature difference to a 10°F temperature difference under minimum load conditions. Design airflow rates shall be sized for the reset supply air temperature, i.e., a 10°F temperature difference.

c. **Direct expansion:** The fuel type for the cooling system shall match that of the cooling system in the proposed design.

d. **VAV:** Constant volume can be modeled if the system qualifies for Exception 1, Section C403.4.5. When the proposed design system has a supply, return or relief fan motor 25 horsepower (hp) or larger, the corresponding fan in the VAV system of the standard reference design shall be modeled assuming a variable speed drive. For smaller fans, a forward-curved centrifugal fan with inlet vanes shall be modeled. If the proposed design's system has a direct digital control system at the zone level, static pressure setpoint reset based on zone requirements in accordance with Section C403.4.2 shall be modeled.

e. **Chilled water:** For systems using purchased chilled water, the chillers are not explicitly modeled and chilled water costs shall be based as determined in Sections C407.3 and C407.5.2. Otherwise, the standard reference design's chiller plant shall be modeled with chillers having the number as indicated in Table C407.5.1(4) as a function of standard reference building chiller plant load and type as indicated in Table C407.5.1(5) as a function of individual chiller load. Where chiller fuel source is mixed, the system in the standard reference design shall have chillers with the same fuel types and with capacities having the same proportional capacity as the proposed design's chillers for each fuel type. Chilled water supply temperature shall be modeled at 44°F design supply temperature and 56°F return temperature. Piping losses shall not be modeled in either building model. Chilled water supply water temperature shall be reset in accordance with Section C403.4.3.4. Pump system power for each pumping system shall be the same as the proposed design; if the proposed design has no chilled water pumps, the standard reference design pump power shall be 22 W/gpm (equal to a pump operating against a 75-foot head, 65-percent combined impeller and motor efficiency). The chilled water system shall be modeled as primary-only variable flow with flow maintained at the design rate through each chiller using a bypass. Chilled water pumps shall be modeled as riding the pump curve or with variable-speed drives when required in Section C403.4.3.4. The heat rejection device shall be an axial fan cooling tower with two-speed fans if required in Section C403.4.4. Condenser water design supply temperature shall be 85°F or 10°F approach to design wet-bulb temperature, whichever is lower, with a design temperature rise of 10°F. The tower shall be controlled to maintain a 70°F leaving water temperature where weather permits, floating up to leaving water temperature at design conditions. Pump system power for each pumping system shall be the same as the proposed design; if the proposed design has no condenser water pumps, the standard reference design pump power shall be 19 W/gpm (equal to a pump operating against a 60-foot head, 60-percent combined impeller and motor efficiency). Each chiller shall be modeled with separate condenser water and chilled water pumps interlocked to operate with the associated chiller.

f. **Fossil fuel boiler:** For systems using purchased hot water or steam, the boilers are not explicitly modeled and hot water or steam costs shall be based on actual utility rates. Otherwise, the boiler plant shall use the same fuel as the proposed design and shall be natural draft. The standard reference design boiler plant shall be modeled with a single boiler if the standard reference design plant load is 600,000 Btu/h and less and with two equally sized boilers for plant capacities exceeding 600,000 Btu/h. Boilers shall be staged as required by the load. Hot water supply temperature shall be modeled at 180°F design supply temperature and 130°F return temperature. Piping losses shall not be modeled in either building model. Hot water supply water temperature shall be reset in accordance with Section C403.4.3.4. Pump system power for each pumping system shall be the same as the proposed design; if the proposed design has no hot water pumps, the standard reference design pump power shall be 19 W/gpm (equal to a pump operating against a 60-foot head, 60-percent combined impeller and motor efficiency). The hot water system shall be modeled as primary only with continuous variable flow. Hot water pumps shall be modeled as riding the pump curve or with variable speed drives when required by Section C403.4.3.4.

g. **Electric heat pump and boiler:** Water-source heat pumps shall be connected to a common heat pump water loop controlled to maintain temperatures between 60°F and 90°F. Heat rejection from the loop shall be provided by an axial fan closed-circuit evaporative fluid cooler with two-speed fans if required in Section C403.4.2. Heat addition to the loop shall be provided by a boiler that uses the same fuel as the proposed design and shall be natural draft. If no boilers exist in the proposed design, the standard reference design building boilers shall be fossil fuel. The standard reference design boiler plant shall be modeled with a single boiler if the standard reference design plant load is 600,000 Btu/h or less and with two equally sized boilers for plant capacities exceeding 600,000 Btu/h. Boilers shall be staged as required by the load. Piping losses shall not be modeled in either building model. Pump system power shall be the same as the proposed design; if the proposed design has no pumps, the standard reference design pump power shall be 22 W/gpm, which is equal to a pump operating against a 75-foot head, with a 65-percent combined impeller and motor efficiency. Loop flow shall be variable with flow shutoff at each heat pump when its compressor cycles off as required by Section C403.4.3.3. Loop pumps shall be modeled as riding the pump curve or with variable speed drives when required by Section C403.4.3.4.

h. **Electric heat pump:** Electric air-source heat pumps shall be modeled with electric auxiliary heat. The system shall be controlled with a multistage space thermostat and an outdoor air thermostat wired to energize auxiliary heat only on the last thermostat stage and when outdoor air temperature is less than 40°F.

i. **Constant volume:** Fans shall be controlled in the same manner as in the proposed design; i.e., fan operation whenever the space is occupied or fan operation cycled on calls for heating and cooling. If the fan is modeled as cycling and the fan energy is included in the energy efficiency rating of the equipment, fan energy shall not be modeled explicitly.

TABLE C407.5.1(4)
NUMBER OF CHILLERS

TOTAL CHILLER PLANT CAPACITY	NUMBER OF CHILLERS
≤ 300 tons	1
> 300 tons, < 600 tons	2, sized equally
≥ 600 tons	2 minimum, with chillers added so that no chiller is larger than 800 tons, all sized equally

For SI: 1 ton = 3517 W.

TABLE C407.5.1(5)
WATER CHILLER TYPES

INDIVIDUAL CHILLER PLANT CAPACITY	ELECTRIC-CHILLER TYPE	FOSSIL FUEL CHILLER TYPE
≤ 100 tons	Reciprocating	Single-effect absorption, direct fired
> 100 tons, < 300 tons	Screw	Double-effect absorption, direct fired
≥ 300 tons	Centrifugal	Double-effect absorption, direct fired

For SI: 1 ton = 3517 W.

C407.6 Calculation software tools. Calculation procedures used to comply with this section shall be software tools capable of calculating the annual energy consumption of all building elements that differ between the *standard reference design* and the *proposed design* and shall include the following capabilities.

1. Computer generation of the *standard reference design* using only the input for the *proposed design.* The calculation procedure shall not allow the user to directly modify the building component characteristics of the *standard reference design.*

2. Building operation for a full calendar year (8,760 hours).

3. Climate data for a full calendar year (8,760 hours) and shall reflect *approved* coincident hourly data for temperature, solar radiation, humidity and wind speed for the building location.

4. Ten or more thermal zones.

5. Thermal mass effects.

6. Hourly variations in occupancy, illumination, receptacle loads, thermostat settings, mechanical ventilation, HVAC equipment availability, service hot water usage and any process loads.

7. Part-load performance curves for mechanical equipment.

8. Capacity and efficiency correction curves for mechanical heating and cooling equipment.

9. Printed *code official* inspection checklist listing each of the *proposed design* component characteristics from Table C407.5.1(1) determined by the analysis to provide compliance, along with their respective performance ratings (e.g., *R*-value, *U*-factor, SHGC, HSPF, AFUE, SEER, EF, etc.).

C407.6.1 Specific approval. Performance analysis tools meeting the applicable subsections of Section C407 and tested according to ASHRAE Standard 140 shall be permitted to be *approved.* Tools are permitted to be *approved* based on meeting a specified threshold for a jurisdiction. The *code official* shall be permitted to approve tools for a specified application or limited scope.

C407.6.2 Input values. Where calculations require input values not specified by Sections C402, C403, C404 and C405, those input values shall be taken from an *approved* source.

SECTION C408
SYSTEM COMMISSIONING

C408.1 General. This section covers the commissioning of the building mechanical systems in Section C403 and electrical power and lighting systems in Section C405.

C408.2 Mechanical systems commissioning and completion requirements. Prior to passing the final mechanical inspection, the *registered design professional* shall provide evidence of mechanical systems *commissioning* and completion in accordance the provisions of this section.

Construction document notes shall clearly indicate provisions for *commissioning* and completion requirements in accordance with this section and are permitted to refer to specifications for further requirements. Copies of all documentation shall be given to the owner and made available to the *code official* upon request in accordance with Sections C408.2.4 and C408.2.5.

Exception: The following systems are exempt from the commissioning requirements:

1. Mechanical systems in buildings where the total mechanical equipment capacity is less than 480,000 Btu/h (140 690 W) cooling capacity and 600,000 Btu/h (175 860 W) heating capacity.

2. Systems included in Section C403.3 that serve dwelling units and sleeping units in hotels, motels, boarding houses or similar units.

C408.2.1 Commissioning plan. A *commissioning plan* shall be developed by a *registered design professional* or approved *agency* and shall include the following items:

1. A narrative description of the activities that will be accomplished during each phase of commissioning, including the personnel intended to accomplish each of the activities.

2. A listing of the specific equipment, appliances or systems to be tested and a description of the tests to be performed.

3. Functions to be tested, including, but not limited to calibrations and economizer controls.

4. Conditions under which the test will be performed. At a minimum, testing shall affirm winter and summer design conditions and full outside air conditions.

5. Measurable criteria for performance.

C408.2.2 Systems adjusting and balancing. HVAC systems shall be balanced in accordance with generally accepted engineering standards. Air and water flow rates shall be measured and adjusted to deliver final flow rates within the tolerances provided in the product specifications. Test and balance activities shall include air system and hydronic system balancing.

** **C408.2.2.1 Air systems balancing.** Each supply air outlet and *zone* terminal device shall be equipped with means for air balancing in accordance with the requirements of Chapter 6 of the *International Mechanical Code*. Discharge dampers are prohibited on constant volume fans and variable volume fans with motors 10 hp (18.6 kW) and larger. Air systems shall be balanced in a manner to first minimize throttling losses then, for fans with system power of greater than 1 hp (0.74 kW), fan speed shall be adjusted to meet design flow conditions.

Exception: Fans with fan motors of 1 hp (0.74 kW) or less.

** **C408.2.2.2 Hydronic systems balancing.** Individual hydronic heating and cooling coils shall be equipped with means for balancing and measuring flow. Hydronic systems shall be proportionately balanced in a manner to first minimize throttling losses, then the pump impeller shall be trimmed or pump speed shall be adjusted to meet design flow conditions. Each hydronic system shall have either the capability to measure pressure across the pump, or test ports at each side of each pump.

Exceptions:

1. Pumps with pump motors of 5 hp (3.7 kW) or less.

2. Where throttling results in no greater than five percent of the nameplate horsepower draw above that required if the impeller were trimmed.

C408.2.3 Functional performance testing. Functional performance testing specified in Sections C408.2.3.1 through C408.2.3.3 shall be conducted.

C408.2.3.1 Equipment. Equipment functional performance testing shall demonstrate the installation and operation of components, systems, and system-to-system interfacing relationships in accordance with approved plans and specifications such that operation, function, and maintenance serviceability for each of the commissioned systems is confirmed. Testing shall include all modes and *sequence of operation*, including under full-load, part-load and the following emergency conditions:

1. All modes as described in the *sequence* of *operation*;

2. Redundant or *automatic* back-up mode;

3. Performance of alarms; and

4. Mode of operation upon a loss of power and restoration of power.

Exception: Unitary or packaged HVAC equipment listed in Tables C403.2.3(1) through C403.2.3(3) that do not require supply air economizers.

C408.2.3.2 Controls. HVAC control systems shall be tested to document that control devices, components, equipment, and systems are calibrated, adjusted and operate in accordance with approved plans and specifications. Sequences of operation shall be functionally tested to document they operate in accordance with *approved* plans and specifications.

C408.2.3.3 Economizers. Air economizers shall undergo a functional test to determine that they operate in accordance with manufacturer's specifications.

C408.2.4 Preliminary commissioning report. A preliminary report of commissioning test procedures and results shall be completed and certified by the *registered design professional* or *approved agency* and provided to the building owner. The report shall be identified as "Preliminary Commissioning Report" and shall identify:

1. Itemization of deficiencies found during testing required by this section that have not been corrected at the time of report preparation.

2. Deferred tests that cannot be performed at the time of report preparation because of climatic conditions.

3. Climatic conditions required for performance of the deferred tests.

C408.2.4.1 Acceptance of report. *Buildings*, or portions thereof, shall not pass the final mechanical inspection until such time as the *code official* has received a letter of transmittal from the *building* owner acknowledging that the *building* owner has received the Preliminary Commissioning Report.

C408.2.4.2 Copy of report. The *code official* shall be permitted to require that a copy of the Preliminary

Commissioning Report be made available for review by the *code official.*

C408.2.5 Documentation requirements. The *construction documents* shall specify that the *documents* described in this section be provided to the *building* owner within 90 days of the date of receipt of the *certificate of occupancy.*

C408.2.5.1 Drawings. Construction documents shall include the location and performance data on each piece of equipment.

C408.2.5.2 Manuals. An operating and maintenance manual shall be provided and include all of the following:

1. Submittal data stating equipment size and selected options for each piece of equipment requiring maintenance.

2. Manufacturer's operation manuals and maintenance manuals for each piece of equipment requiring maintenance, except equipment not furnished as part of the project. Required routine maintenance actions shall be clearly identified.

3. Name and address of at least one service agency.

4. HVAC controls system maintenance and calibration information, including wiring diagrams, schematics, and control sequence descriptions. Desired or field-determined setpoints shall be permanently recorded on control drawings at control devices or, for digital control systems, in system programming instructions.

5. A narrative of how each system is intended to operate, including recommended setpoints.

C408.2.5.3 System balancing report. A written report describing the activities and measurements completed in accordance with Section C408.2.2.

C408.2.5.4 Final commissioning report. A report of test procedures and results identified as "Final Commissioning Report" shall be delivered to the building owner and shall include:

1. Results of functional performance tests.

2. Disposition of deficiencies found during testing, including details of corrective measures used or proposed.

3. Functional performance test procedures used during the commissioning process including measurable criteria for test acceptance, provided herein for repeatability.

Exception: Deferred tests which cannot be performed at the time of report preparation due to climatic conditions.

C408.3 Lighting system functional testing. Controls for automatic lighting systems shall comply with Section C408.3.

C408.3.1 Functional testing. Testing shall ensure that control hardware and software are calibrated, adjusted, programmed and in proper working condition in accor-

dance with the construction documents and manufacturer's installation instructions. The construction documents shall state the party who will conduct the required functional testing. Where required by the code official, an approved party independent from the design or construction of the project shall be responsible for the functional testing and shall provide documentation to the code official certifying that the installed lighting controls meet the provisions of Section C405.

Where occupant sensors, time switches, programmable schedule controls, photosensors or daylighting controls are installed, the following procedures shall be performed:

1. Confirm that the placement, sensitivity and time-out adjustments for occupant sensors yield acceptable performance.

2. Confirm that the time switches and programmable schedule controls are programmed to turn the lights off.

3. Confirm that the placement and sensitivity adjustments for photosensor controls reduce electric light based on the amount of usable daylight in the space as specified.

CHAPTER 5 [CE]

REFERENCED STANDARDS

This chapter lists the standards that are referenced in various sections of this document. The standards are listed herein by the promulgating agency of the standard, the standard identification, the effective date and title, and the section or sections of this document that reference the standard. The application of the referenced standards shall be as specified in Section C106.

AAMA

American Architectural Manufacturers Association
1827 Walden Office Square
Suite 550
Schaumburg, IL 60173-4268

Standard reference number	Title	Referenced in code section number
AAMA/WDMA/CSA 101/I.S.2/A C440—11	North American Fenestration Standard/ Specifications for Windows, Doors and Unit Skylights	Table C402.4.3

AHAM

Association of Home Appliance Manufacturers
1111 19th Street, NW, Suite 402
Washington, DC 20036

Standard reference number	Title	Referenced in code section number
ANSI/ AHAM RAC-1—2008	Room Air Conditioners	Table C403.2.3(3)

AHRI

Air Conditioning, Heating, and Refrigeration Institute
4100 North Fairfax Drive
Suite 200
Arlington, VA 22203

Standard reference number	Title	Referenced in code section number
ISO/AHRI/ASHRAE 13256-1 (2005)	Water-source Heat Pumps—Testing and Rating for Performance— Part 1: Water-to-air and Brine-to-air Heat Pumps	Table C403.2.3(2)
ISO/AHRI/ASHRAE 13256-2 (1998)	Water-source Heat Pumps—Testing and Rating for Performance— Part 2: Water-to-water and Brine-to-water Heat Pumps	Table C403.2.3(2)
210/240—08	Unitary Air Conditioning and Air-source Heat Pump Equipment	Table C403.2.3(1), Table C403.2.3(2)
310/380—04	Standard for Packaged Terminal Air Conditioners and Heat Pumps	Table C403.2.3(3)
340/360—2007	Commercial and Industrial Unitary Air-conditioning and Heat Pump Equipment	Table C403.2.3(1), Table C403.2.3(2)
365—09	Commercial and Industrial Unitary Air-conditioning Condensing Units	Table C403.2.3(1), Table C403.2.3(6)
390—03	Performance Rating of Single Package Vertical Air Conditioners and Heat Pumps	Table C403.2.3(3)
400—01	Liquid to Liquid Heat Exchangers with Addendum 2	Table C403.2.3(9)
440—08	Room Fan Coil	C403.2.8
460—05	Performance Rating Remote Mechanical Draft Air-cooled Refrigerant Condensers	Table C403.2.3(8)
550/590—03	Water Chilling Packages Using the Vapor Compression Cycle—with Addenda	C403.2.3.1, Table C403.2.3(7), Table C406.2(6)
560—00	Absorption Water Chilling and Water-heating Packages	Table C403.2.3(7)
1160—08	Performance Rating of Heat Pump Pool Heaters	Table C404.2

AMCA

Air Movement and Control Association International
30 West University Drive
Arlington Heights, IL 60004-1806

Standard reference number	Title	Referenced in code section number
500D—10	Laboratory Methods for Testing Dampers for Rating	C402.4.5.1, C402.4.5.2

ANSI

American National Standards Institute
25 West 43rd Street
Fourth Floor
New York, NY 10036

Standard reference number	Title	Referenced in code section number
Z21.10.3/CSA 4.3—04	Gas Water Heaters, Volume III—Storage Water Heaters with Input Ratings Above 75,000 Btu per Hour, Circulating Tank and Instantaneous	Table C404.2
Z21.47/CSA 2.3—06	Gas-fired Central Furnaces	Table C403.2.3(4), Table C406.2.(4)
Z83.8/CSA 2.6—09	Gas Unit Heaters, Gas Packaged Heaters, Gas Utility Heaters and Gas-fired Duct Furnaces	Table C403.2.3(4), Table C406.2.(4)

ASHRAE

American Society of Heating, Refrigerating and Air-Conditioning Engineers, Inc.
1791 Tullie Circle, NE
Atlanta, GA 30329-2305

Standard reference number	Title	Referenced in code section number
ANSI/ASHRAE/ACCA Standard 183—2007	Peak Cooling and Heating Load Calculations in Buildings, Except Low-rise Residential Buildings	C403.2.1
ASHRAE—2004	ASHRAE HVAC Systems and Equipment Handbook—2004	C403.2.1
ISO/AHRI/ASHRAE 13256-1 (2005)	Water-source Heat Pumps—Testing and Rating for Performance— Part 1: Water-to-air and Brine-to-air Heat Pumps	Table C403.2.3(2)
ISO/AHRI/ASHRAE 13256-2 (1998)	Water-source Heat Pumps—Testing and Rating for Performance— Part 2: Water-to-water and Brine-to-water Heat Pumps	Table C403.2.3(2)
90.1—2010	Energy Standard for Buildings Except Low-rise Residential Buildings	C401.2, C401.2.1, C402.1.1, Table C402.1.2, Table C402.2, Table C407.6.1
119—88 (RA 2004)	Air Leakage Performance for Detached Single-family Residential Buildings	Table C405.5.2(1)
140—2010	Standard Method of Test for the Evaluation of Building Energy Analysis Computer Programs	C407.6.1
146—2006	Testing and Rating Pool Heaters	Table C404.2

ASTM

ASTM International
100 Barr Harbor Drive
West Conshohocken, PA 19428-2859

Standard reference number	Title	Referenced in code section number
C 90—08	Specification for Load-bearing Concrete Masonry Units	Table C402.2
C 1371—04	Standard Test Method for Determination of Emittance of Materials Near Room Temperature Using Portable Emissometers	Table C402.2.1.1
C 1549—04	Standard Test Method for Determination of Solar Reflectance Near Ambient Temperature Using A Portable Solar Reflectometer	Table C405.2.1.1
D 1003—07e1	Standard Test Method for Haze and Luminous Transmittance of Transparent Plastics	C402.3.2.2

ASTM—continued

E 283—04	Test Method for Determining the Rate of Air Leakage Through Exterior Windows, Curtain Walls and Doors Under Specified Pressure Differences Across the Specimen . . . . Table C402.2.1.1, C402.4.1.2.2, Table C402.4.3, C402.4.4, C402.4.8
E 408—71(2002)	Test Methods for Total Normal Emittance of Surfaces Using Inspection-meter Techniques . . . . Table C402.2.1.1
E 779—03	Standard Test Method for Determining Air Leakage Rate by Fan Pressurization C402.4.1.2.3
E 903—96	Standard Test Method Solar Absorptance, Reflectance and Transmittance of Materials Using Integrating Spheres (Withdrawn 2005) . . . . Table C402.2.1.1
E 1677—05	Standard Specification for an Air-retarder (AR) Material or System for Low-rise Framed Building Walls C402.4.1.2.2
E 1918—97	Standard Test Method for Measuring Solar Reflectance of Horizontal or Low-sloped Surfaces in the Field . . . . Table C402.2.1.1
E 1980—(2001)	Standard Practice for Calculating Solar Reflectance Index of Horizontal and Low-sloped Opaque Surfaces . . . . Table C402.2.1.1
E 2178—03	Standard Test Method for Air Permanence of Building Materials C402.4.1.2.1
E 2357—05	Standard Test Method for Determining Air Leakage of Air Barriers Assemblies C404.1.2.2

CSA

Canadian Standards Association
5060 Spectrum Way
Mississauga, Ontario, Canada L4W 5N6

Standard reference number	Title	Referenced in code section number
AAMA/WDMA/CSA 101/I.S.2/A440—11	North American Fenestration Standard/Specification for Windows, Doors and Unit Skylights	Table C402.4.3

CTI

Cooling Technology Institute
2611 FM 1960 West, Suite A-101
Houston, TX 77068

Standard reference number	Title	Referenced in code section number
ATC 105 (00)	Acceptance Test Code for Water Cooling Tower	Table C403.2.3(8)
STD 201—09	Standard for Certification of Water Cooling Towers Thermal Performances	Table C403.2.3(8)

DASMA

Door and Access Systems Manufacturers Association
1300 Sumner Avenue
Cleveland, OH 44115-2851

Standard reference number	Title	Referenced in code section number
105—92 (R2004)	Test Method for Thermal Transmittance and Air Infiltration of Garage Doors	Table C402.4.3

DOE

U.S. Department of Energy
c/o Superintendent of Documents
U.S. Government Printing Office
Washington, DC 20402-9325

Standard reference number	Title	Referenced in code section number
10 CFR, Part 430—1998	Energy Conservation Program for Consumer Products: Test Procedures and Certification and Enforcement Requirement for Plumbing Products; and Certification and Enforcement Requirements for Residential Appliances; Final Rule	Table C403.2.3(4), Table C403.2.3(5), Table C404.2, Table C406.2(4), Table C406.2(5)
10 CFR, Part 430, Subpart B, Appendix N—1998	Uniform Test Method for Measuring the Energy Consumption of Furnaces and Boilers	C202
10 CFR, Part 431—2004	Energy Efficiency Program for Certain Commercial and Industrial Equipment: Test Procedures and Efficiency Standards; Final Rules	Table C403.2.3(5), Table C406.2(5)
NAECA 87—(88)	National Appliance Energy Conservation Act 1987 [(Public Law 100-12 (with Amendments of 1988-P.L. 100-357)]	Tables C403.2.3(1), (2), (4)

ICC

International Code Council, Inc.
500 New Jersey Avenue, NW
6th Floor
Washington, DC 20001

Standard reference number	Title	Referenced in code section number
IBC—12	International Building Code®	C201.3, C303.2, C402.4.4
IFC—12	International Fire Code®	C201.3
IFGC—12	International Fuel Gas Code®	C201.3
IMC—12	International Mechanical Code®	C403.2.5, C403.2.5.1, C403.2.6, C403.2.7, C403.2.7.1, C403.2.7.1.1, C403.2.7.1.2, C403.2.7.1.3, C403.4.5, C408.2.2.1
IPC—12	International Plumbing Code®	C201.3

IESNA

Illuminating Engineering Society of North America
120 Wall Street, 17th Floor
New York, NY 10005-4001

Standard reference number	Title	Referenced in code section number
ANSI/ASHRAE/IESNA 90.1—2010	Energy Standard for Buildings, Except Low-rise Residential Buildings	C401.2, C401.2.1, C402.1.1, Table C402.1.2, Table C402.2

ISO

International Organization for Standardization
1, rue de Varembe, Case postale 56, CH-1211
Geneva, Switzerland

Standard reference number	Title	Referenced in code section number
ISO/AHRI/ASHRAE 13256-1 (2005)	Water-source Heat Pumps—Testing and Rating for Performance— Part 1: Water-to-air and Brine-to-air Heat Pumps	C403.2.3(2)
ISO/AHRI/ASHRAE 13256-2 (1998)	Water-Source Heat Pumps—Testing and Rating for Performance— Part 2: Water-to-water and Brine-to-water Heat Pumps	C403.2.3(2)

NFRC

National Fenestration Rating Council, Inc.
6305 Ivy Lane, Suite 140
Greenbelt, MD 20770

Standard reference number	Title	Referenced in code section number
100—2010	Procedure for Determining Fenestration Products U-factors	C303.1.2, C402.2.1
200—2010	Procedure for Determining Fenestration Product Solar Heat Gain Coefficients and Visible Transmittance at Normal Incidence	C303.1.3, C402.3.1.1
400—2010	Procedure for Determining Fenestration Product Air Leakage	Table C402.4.3

SMACNA

Sheet Metal and Air Conditioning Contractors National Association, Inc.
4021 Lafayette Center Drive
Chantilly, VA 20151-1209

Standard reference number	Title	Referenced in code section number
SMACNA—85	HVAC Air Duct Leakage Test Manual	C403.2.7.1.3

UL

Underwriters Laboratories
333 Pfingsten Road
Northbrook, IL 60062-2096

Standard reference number	Title	Referenced in code section number
727—06	Oil-fired Central Furnaces—with Revisions through April 2010	Table C403.2.3(4), Table C406.2(4)
731—95	Oil-fired Unit Heaters—with Revisions through April 2010	Table C403.2.3(4), Table C406.2(4)

US-FTC

United States-Federal Trade Commission
600 Pennsylvania Avenue NW
Washington, DC 20580

Standard reference number	Title	Referenced in code section number
CFR Title 16 (May 31, 2005)	R-value Rule	C303.1.4

WDMA

Window and Door Manufacturers Association
1400 East Touhy Avenue, Suite 470
Des Plaines, IL 60018

Standard reference number	Title	Referenced in code section number
AAMA/WDMA/CSA 101/I.S.2/A440—11	North American Fenestration Standard/Specification for Windows, Doors and Unit Skylights	Table C402.4.3

INDEX [CE]

E

F

W

Z

EDITORIAL CHANGES – SECOND PRINTING

Page C-30, Section C402.2.5: item 2, line 2 now reads . . . 120 pcf (1,900 kg/m^3).

Page C-34, Table C402.3.3.1: column 1, row 3 now reads . . . PF $\geq$ 0.5

Page C-39, Table C403.2.3(1): column 5, row 10, line 1 now reads . . . 11.0 EER

Page C-39, Table C403.2.3(1): column 5, row 10, line 2 now reads . . . 11.1 EER

Page C-39, Table C403.2.3(1): column 6, row 10, line 2 now reads . . . 11.9 EER

Page C-39, Table C403.2.3(1): column 5, row 11, line 2 now reads . . . 10.9 EER

Page C-39, Table C403.2.3(1): column 6, row 11, line 2 now reads . . . 11.7 EER

Page C-42, Table C403.2.3(3): column 4, row 7 now reads . . . 3.2 - (0.026 × Cap/1000) COP

Page C-42, Table C403.2.3(3): column 5, row 7 now reads . . . 3.2 - (0.026 × Cap/1000) COP

Page C-42, Table C403.2.3(3): column 4, row 8 now reads . . . 2.9 - (0.026 × Cap/1000) COP

Page C-42, Table C403.2.3(3): column 5, row 8 now reads . . . 2.9 - (0.026 × Cap/1000) COP

Page C-55, Section C403.4.5: line 3 now reads . . . through C403.4.5.4 shall apply to complex mechanical

Page C-67, Table C406.2(6): column 8, row 3 now reads . . . AHRI 550/590^f

Page C-80, IESNA Referenced Standard: first column now reads . . . ANSI/ASHRAE/IESNA 90.1—2010

EDITORIAL CHANGES – THIRD PRINTING

Page C-5, Section C106.1.1, line 1 now reads . . . C106.1.1 Conflicts. Where conflicts occur between provi-

Page C-7, COEFFICIENT OF PERFORMANCE (COP) – HEATING now reads . . . COEFFICIENT OF PERFORMANCE (COP) – HEATING. The ratio of the rate of heat delivered to the rate of energy input, in consistent units, for a complete heat pump system, including the compressor and, if applicable, auxiliary heat, under designated operating conditions.

Page C-32, Table C402.2: column 2, row 8 now reads . . . R-5.7ci^c

Page C-32, Table C402.2: column 3, row 8 now reads . . . R-5.7ci^c

Page C-32, Table C402.2: column 4, row 8 now reads . . . R-5.7ci^c

Page C-32, TABLE C402.2: column 1, row 7, cell 1, now reads . . . Mass^c

Page C-35, Section C402.4.1.1: item 3, line 2 now reads . . . tion C402.4.8. Where similar objects are

Page C-36, Section C402.4.4: Exception, row 2 now reads . . . Section 716 or 716.4 of the *International Building*

Page C-37, Section C402.4.8: line 5 now reads . . . an air leakage rate of not more than 2.0 cfm (0.944 L/s) when

Page C-38, TABLE C403.2.3(1): column 5, Before 6/1/2011, row 21, line 2 now reads . . . 11.1 EER

Page C-38, TABLE C403.2.3(1): column 5, Before 6/1/2011, row 22, line 2 now reads . . . 10.9 EER

Page C-38, TABLE C403.2.3(1): column 5, Before 6/1/2011, row 23, line 2 now reads . . . 11.1 EER

Page C-38, TABLE C403.2.3(1): column 5, Before 6/1/2011, row 24, line 2 now reads . . . 10.9 EER

Page C-38, TABLE C403.2.3(1): column 5, As of 6/1/2011 row 21, line 2 now reads . . . 12.6 EER

Page C-38, TABLE C403.2.3(1): column 5, As of 6/1/2011 row 22, line 2 now reads . . . 12.4 EER

Page C-38, TABLE C403.2.3(1): column 5, As of 6/1/2011 row 23, line 2 now reads . . . 12.4 EER

Page C-38, TABLE C403.2.3(1): column 5, As of 6/1/2011 row 24, line 2 now reads . . . 12.2 EER

Page C-39, TABLE C403.2.3(1)—continued: column 5, Before 6/1/2011, row 8, line 2 now reads . . . 11.1 EER

Page C-39, TABLE C403.2.3(1)—continued: column 5, Before 6/1/2011, row 9, line 2 now reads . . . 10.9 EER

Page C-39, TABLE C403.2.3(1)—continued: column 5, As of 6/1/2011, row 8, line 2 now reads . . . 12.1 EER

Page C-39, TABLE C403.2.3(1)—continued: column 5, As of 6/1/2011, row 9, line 2 now reads . . . 11.9 EER

Page C-43, TABLE C403.2.3(3)—continued: column 1, row 3, cell 1 now reads . . . Room air conditioners, without louvered slides

Page C-61, Sections C405.5.1.1 through C405.5.1.4 has been deleted.

Page C-70, TABLE C407.5.1(1): column 2, row 11, line 2 now reads . . . Section C405.5.2. Where the occupancy of the building is not known,

Page C-81, NFRC: standard reference number 100—2009 now reads . . . 100—2010

Page C-81, NFRC: standard reference number 200—2009 now reads . . . 200—2010

Page C-81, NFRC: standard reference number 400—2009 now reads . . . 400—2010

EDITORIAL CHANGES – FOURTH PRINTING

Page C-61, Section C405.5.1.1 has been added.

Page C-61, Section C405.5.1.2 has been added.

Page C-61, Section C405.5.1.3 has been added.

Page C-61, Section C405.5.1.4 has been added.

Page C-67, Table C406.2(6): row 1, column 4 now reads . . . MINIMUM EFFICIENCY[c] (I-P)

EDITORIAL CHANGES – FIFTH PRINTING

Page C-36, Section C402.4.4: Exception line 2 now reads . . . Section 716 or 716.5 of the *International Building*

IECC—RESIDENTIAL PROVISIONS

TABLE OF CONTENTS

CHAPTER 1 [RE]

SCOPE AND ADMINISTRATION

PART 1—SCOPE AND APPLICATION

SECTION R101
SCOPE AND GENERAL REQUIREMENTS

R101.1 Title. This code shall be known as the *International Energy Conservation Code* of **[NAME OF JURISDICTION]**, and shall be cited as such. It is referred to herein as "this code."

R101.2 Scope. This code applies to *residential buildings* and the buildings sites and associated systems and equipment.

R101.3 Intent. This code shall regulate the design and construction of buildings for the effective use and conservation of energy over the useful life of each building. This code is intended to provide flexibility to permit the use of innovative approaches and techniques to achieve this objective. This code is not intended to abridge safety, health or environmental requirements contained in other applicable codes or ordinances.

R101.4 Applicability. Where, in any specific case, different sections of this code specify different materials, methods of construction or other requirements, the most restrictive shall govern. Where there is a conflict between a general requirement and a specific requirement, the specific requirement shall govern.

R101.4.1 Existing buildings. Except as specified in this chapter, this code shall not be used to require the removal, *alteration* or abandonment of, nor prevent the continued use and maintenance of, an existing building or building system lawfully in existence at the time of adoption of this code.

R101.4.2 Historic buildings. Any building or structure that is listed in the State or National Register of Historic Places; designated as a historic property under local or state designation law or survey; certified as a contributing resource with a National Register listed or locally designated historic district; or with an opinion or certification that the property is eligible to be listed on the National or State Registers of Historic Places either individually or as a contributing building to a historic district by the State Historic Preservation Officer or the Keeper of the National Register of Historic Places, are exempt from this code.

R101.4.3 Additions, alterations, renovations or repairs. Additions, alterations, renovations or repairs to an existing building, building system or portion thereof shall conform to the provisions of this code as they relate to new construction without requiring the unaltered portion(s) of the existing building or building system to comply with this code. Additions, alterations, renovations or repairs shall not create an unsafe or hazardous condition or overload existing building systems. An addition shall be deemed to comply with this code if the addition alone complies or if the existing building and addition comply with this code as a single building.

Exception: The following need not comply provided the energy use of the building is not increased:

1. Storm windows installed over existing fenestration.

2. Glass only replacements in an existing sash and frame.

3. Existing ceiling, wall or floor cavities exposed during construction provided that these cavities are filled with insulation.

4. Construction where the existing roof, wall or floor cavity is not exposed.

5. Reroofing for roofs where neither the sheathing nor the insulation is exposed. Roofs without insulation in the cavity and where the sheathing or insulation is exposed during reroofing shall be insulated either above or below the sheathing.

6. Replacement of existing doors that separate *conditioned space* from the exterior shall not require the installation of a vestibule or revolving door, provided, however, that an existing vestibule that separates a *conditioned space* from the exterior shall not be removed,

7. Alterations that replace less than 50 percent of the luminaires in a space, provided that such alterations do not increase the installed interior lighting power.

8. Alterations that replace only the bulb and ballast within the existing luminaires in a space provided that the *alteration* does not increase the installed interior lighting power.

R101.4.4 Change in occupancy or use. Spaces undergoing a change in occupancy that would result in an increase in demand for either fossil fuel or electrical energy shall comply with this code.

R101.4.5 Change in space conditioning. Any nonconditioned space that is altered to become *conditioned space* shall be required to be brought into full compliance with this code.

R101.4.6 Mixed occupancy. Where a building includes both *residential* and *commercial* occupancies, each occupancy shall be separately considered and meet the applicable provisions of the IECC—Commercial and Residential Provisions.

R101.5 Compliance. *Residential buildings* shall meet the provisions of IECC—Residential Provisions. *Commercial buildings* shall meet the provisions of IECC—Commercial Provisions.

R101.5.1 Compliance materials. The *code official* shall be permitted to approve specific computer software, worksheets, compliance manuals and other similar materials that meet the intent of this code.

R101.5.2 Low energy buildings. The following buildings, or portions thereof, separated from the remainder of the building by *building thermal envelope* assemblies complying with this code shall be exempt from the *building thermal envelope* provisions of this code:

1. Those with a peak design rate of energy usage less than 3.4 Btu/h · ft^2 (10.7 W/m^2) or 1.0 watt/ft^2 (10.7 W/m^2) of floor area for space conditioning purposes.

2. Those that do not contain *conditioned space*.

SECTION R102
ALTERNATE MATERIALS—METHOD OF CONSTRUCTION, DESIGN OR INSULATING SYSTEMS

R102.1 General. This code is not intended to prevent the use of any material, method of construction, design or insulating system not specifically prescribed herein, provided that such construction, design or insulating system has been *approved* by the *code official* as meeting the intent of this code.

R102.1.1 Above code programs. The *code official* or other authority having jurisdiction shall be permitted to deem a national, state or local energy efficiency program to exceed the energy efficiency required by this code. Buildings *approved* in writing by such an energy efficiency program shall be considered in compliance with this code. The requirements identified as "mandatory" in Chapter 4 shall be met.

PART 2—ADMINISTRATION AND ENFORCEMENT

SECTION R103
CONSTRUCTION DOCUMENTS

R103.1 General. Construction documents and other supporting data shall be submitted in one or more sets with each application for a permit. The construction documents shall be prepared by a registered design professional where required by the statutes of the jurisdiction in which the project is to be constructed. Where special conditions exist, the *code official* is authorized to require necessary construction documents to be prepared by a registered design professional.

Exception: The *code official* is authorized to waive the requirements for construction documents or other supporting data if the *code official* determines they are not necessary to confirm compliance with this code.

R103.2 Information on construction documents. Construction documents shall be drawn to scale upon suitable material. Electronic media documents are permitted to be submitted when *approved* by the *code official*. Construction documents shall be of sufficient clarity to indicate the loca-

tion, nature and extent of the work proposed, and show in sufficient detail pertinent data and features of the building, systems and equipment as herein governed. Details shall include, but are not limited to, as applicable, insulation materials and their *R*-values; fenestration *U*-factors and SHGCs; area-weighted *U*-factor and SHGC calculations; mechanical system design criteria; mechanical and service water heating system and equipment types, sizes and efficiencies; economizer description; equipment and systems controls; fan motor horsepower (hp) and controls; duct sealing, duct and pipe insulation and location; lighting fixture schedule with wattage and control narrative; and air sealing details.

R103.3 Examination of documents. The *code official* shall examine or cause to be examined the accompanying construction documents and shall ascertain whether the construction indicated and described is in accordance with the requirements of this code and other pertinent laws or ordinances.

R103.3.1 Approval of construction documents. When the *code official* issues a permit where construction documents are required, the construction documents shall be endorsed in writing and stamped "Reviewed for Code Compliance." Such *approved* construction documents shall not be changed, modified or altered without authorization from the *code official*. Work shall be done in accordance with the *approved* construction documents.

One set of construction documents so reviewed shall be retained by the *code official*. The other set shall be returned to the applicant, kept at the site of work and shall be open to inspection by the *code official* or a duly authorized representative.

R103.3.2 Previous approvals. This code shall not require changes in the construction documents, construction or designated occupancy of a structure for which a lawful permit has been heretofore issued or otherwise lawfully authorized, and the construction of which has been pursued in good faith within 180 days after the effective date of this code and has not been abandoned.

R103.3.3 Phased approval. The *code official* shall have the authority to issue a permit for the construction of part of an energy conservation system before the construction documents for the entire system have been submitted or *approved*, provided adequate information and detailed statements have been filed complying with all pertinent requirements of this code. The holders of such permit shall proceed at their own risk without assurance that the permit for the entire energy conservation system will be granted.

R103.4 Amended construction documents. Changes made during construction that are not in compliance with the *approved* construction documents shall be resubmitted for approval as an amended set of construction documents.

R103.5 Retention of construction documents. One set of *approved* construction documents shall be retained by the *code official* for a period of not less than 180 days from date of completion of the permitted work, or as required by state or local laws.

SECTION R104
INSPECTIONS

R104.1 General. Construction or work for which a permit is required shall be subject to inspection by the *code official*.

R104.2 Required approvals. Work shall not be done beyond the point indicated in each successive inspection without first obtaining the approval of the *code official*. The *code official*, upon notification, shall make the requested inspections and shall either indicate the portion of the construction that is satisfactory as completed, or notify the permit holder or his or her agent wherein the same fails to comply with this code. Any portions that do not comply shall be corrected and such portion shall not be covered or concealed until authorized by the *code official*.

R104.3 Final inspection. The building shall have a final inspection and not be occupied until *approved*.

R104.4 Reinspection. A building shall be reinspected when determined necessary by the *code official*.

R104.5 Approved inspection agencies. The *code official* is authorized to accept reports of *approved* inspection agencies, provided such agencies satisfy the requirements as to qualifications and reliability.

R104.6 Inspection requests. It shall be the duty of the holder of the permit or their duly authorized agent to notify the *code official* when work is ready for inspection. It shall be the duty of the permit holder to provide access to and means for inspections of such work that are required by this code.

R104.7 Reinspection and testing. Where any work or installation does not pass an initial test or inspection, the necessary corrections shall be made so as to achieve compliance with this code. The work or installation shall then be resubmitted to the *code official* for inspection and testing.

R104.8 Approval. After the prescribed tests and inspections indicate that the work complies in all respects with this code, a notice of approval shall be issued by the *code official*.

R104.8.1 Revocation. The *code official* is authorized to, in writing, suspend or revoke a notice of approval issued under the provisions of this code wherever the certificate is issued in error, or on the basis of incorrect information supplied, or where it is determined that the building or structure, premise, or portion thereof is in violation of any ordinance or regulation or any of the provisions of this code.

SECTION R105
VALIDITY

R105.1 General. If a portion of this code is held to be illegal or void, such a decision shall not affect the validity of the remainder of this code.

SECTION R106
REFERENCED STANDARDS

R106.1 Referenced codes and standards. The codes and standards referenced in this code shall be those listed in Chapter 5, and such codes and standards shall be considered as part of the requirements of this code to the prescribed extent of each such reference and as further regulated in Sections R106.1.1 and R106.1.2.

R106.1.1 Conflicts. Where conflicts occur between provisions of this code and referenced codes and standards, the provisions of this code shall apply.

R106.1.2 Provisions in referenced codes and standards. Where the extent of the reference to a referenced code or standard includes subject matter that is within the scope of this code, the provisions of this code, as applicable, shall take precedence over the provisions in the referenced code or standard.

R106.2 Conflicting requirements. Where the provisions of this code and the referenced standards conflict, the provisions of this code shall take precedence.

R106.3 Application of references. References to chapter or section numbers, or to provisions not specifically identified by number, shall be construed to refer to such chapter, section or provision of this code.

R106.4 Other laws. The provisions of this code shall not be deemed to nullify any provisions of local, state or federal law.

SECTION R107
FEES

R107.1 Fees. A permit shall not be issued until the fees prescribed in Section R107.2 have been paid, nor shall an amendment to a permit be released until the additional fee, if any, has been paid.

R107.2 Schedule of permit fees. A fee for each permit shall be paid as required, in accordance with the schedule as established by the applicable governing authority.

R107.3 Work commencing before permit issuance. Any person who commences any work before obtaining the necessary permits shall be subject to an additional fee established by the *code official*, which shall be in addition to the required permit fees.

R107.4 Related fees. The payment of the fee for the construction, *alteration*, removal or demolition of work done in connection to or concurrently with the work or activity authorized by a permit shall not relieve the applicant or holder of the permit from the payment of other fees that are prescribed by law.

R107.5 Refunds. The *code official* is authorized to establish a refund policy.

SECTION R108
STOP WORK ORDER

R108.1 Authority. Whenever the *code official* finds any work regulated by this code being performed in a manner either contrary to the provisions of this code or dangerous or unsafe, the *code official* is authorized to issue a stop work order.

R108.2 Issuance. The stop work order shall be in writing and shall be given to the owner of the property involved, or to the owner's agent, or to the person doing the work. Upon issuance of a stop work order, the cited work shall immediately cease. The stop work order shall state the reason for the order, and the conditions under which the cited work will be permitted to resume.

R108.3 Emergencies. Where an emergency exists, the *code official* shall not be required to give a written notice prior to stopping the work.

R108.4 Failure to comply. Any person who shall continue any work after having been served with a stop work order, except such work as that person is directed to perform to remove a violation or unsafe condition, shall be liable to a fine of not less than **[AMOUNT]** dollars or more than **[AMOUNT]** dollars.

SECTION R109
BOARD OF APPEALS

R109.1 General. In order to hear and decide appeals of orders, decisions or determinations made by the *code official* relative to the application and interpretation of this code, there shall be and is hereby created a board of appeals. The *code official* shall be an ex officio member of said board but shall have no vote on any matter before the board. The board of appeals shall be appointed by the governing body and shall hold office at its pleasure. The board shall adopt rules of procedure for conducting its business, and shall render all decisions and findings in writing to the appellant with a duplicate copy to the *code official*.

R109.2 Limitations on authority. An application for appeal shall be based on a claim that the true intent of this code or the rules legally adopted thereunder have been incorrectly interpreted, the provisions of this code do not fully apply or an equally good or better form of construction is proposed. The board shall have no authority to waive requirements of this code.

R109.3 Qualifications. The board of appeals shall consist of members who are qualified by experience and training and are not employees of the jurisdiction.

DEFINITIONS

SECTION R201
GENERAL

R201.1 Scope. Unless stated otherwise, the following words and terms in this code shall have the meanings indicated in this chapter.

R201.2 Interchangeability. Words used in the present tense include the future; words in the masculine gender include the feminine and neuter; the singular number includes the plural and the plural includes the singular.

R201.3 Terms defined in other codes. Terms that are not defined in this code but are defined in the *International Building Code, International Fire Code, International Fuel Gas Code, International Mechanical Code, International Plumbing Code* or the *International Residential Code* shall have the meanings ascribed to them in those codes.

R201.4 Terms not defined. Terms not defined by this chapter shall have ordinarily accepted meanings such as the context implies.

SECTION R202
GENERAL DEFINITIONS

ABOVE-GRADE WALL. A wall more than 50 percent above grade and enclosing *conditioned space*. This includes between-floor spandrels, peripheral edges of floors, roof and basement knee walls, dormer walls, gable end walls, walls enclosing a mansard roof and skylight shafts.

ACCESSIBLE. Admitting close approach as a result of not being guarded by locked doors, elevation or other effective means (see "Readily *accessible*").

ADDITION. An extension or increase in the *conditioned space* floor area or height of a building or structure.

AIR BARRIER. Material(s) assembled and joined together to provide a barrier to air leakage through the building envelope. An air barrier may be a single material or a combination of materials.

ALTERATION. Any construction or renovation to an existing structure other than repair or addition that requires a permit. Also, a change in a mechanical system that involves an extension, addition or change to the arrangement, type or purpose of the original installation that requires a permit.

APPROVED. Approval by the *code official* as a result of investigation and tests conducted by him or her, or by reason of accepted principles or tests by nationally recognized organizations.

AUTOMATIC. Self-acting, operating by its own mechanism when actuated by some impersonal influence, as, for example, a change in current strength, pressure, temperature or mechanical configuration (see "Manual").

BASEMENT WALL. A wall 50 percent or more below grade and enclosing *conditioned space*.

BUILDING. Any structure used or intended for supporting or sheltering any use or occupancy, including any mechanical systems, service water heating systems and electric power and lighting systems located on the building site and supporting the building.

BUILDING SITE. A contiguous area of land that is under the ownership or control of one entity.

BUILDING THERMAL ENVELOPE. The basement walls, exterior walls, floor, roof, and any other building elements that enclose *conditioned space* or provides a boundary between *conditioned space* and exempt or unconditioned space.

C-FACTOR (THERMAL CONDUCTANCE). The coefficient of heat transmission (surface to surface) through a building component or assembly, equal to the time rate of heat flow per unit area and the unit temperature difference between the warm side and cold side surfaces (Btu/h ft^2 × °F) [W/(m^2 × K)].

CODE OFFICIAL. The officer or other designated authority charged with the administration and enforcement of this code, or a duly authorized representative.

COMMERCIAL BUILDING. For this code, all buildings that are not included in the definition of "Residential buildings."

CONDITIONED FLOOR AREA. The horizontal projection of the floors associated with the *conditioned space*.

CONDITIONED SPACE. An area or room within a building being heated or cooled, containing uninsulated ducts, or with a fixed opening directly into an adjacent *conditioned space*.

CONTINUOUS AIR BARRIER. A combination of materials and assemblies that restrict or prevent the passage of air through the building thermal envelope.

CRAWL SPACE WALL. The opaque portion of a wall that encloses a crawl space and is partially or totally below grade.

CURTAIN WALL. Fenestration products used to create an external nonload-bearing wall that is designed to separate the exterior and interior environments.

DEMAND RECIRCULATION WATER SYSTEM. A water distribution system where pump(s) prime the service hot water piping with heated water upon demand for hot water.

DUCT. A tube or conduit utilized for conveying air. The air passages of self-contained systems are not to be construed as air ducts.

DUCT SYSTEM. A continuous passageway for the transmission of air that, in addition to ducts, includes duct fittings, dampers, plenums, fans and accessory air-handling equipment and appliances.

DWELLING UNIT. A single unit providing complete independent living facilities for one or more persons, including permanent provisions for living, sleeping, eating, cooking and sanitation.

ENERGY ANALYSIS. A method for estimating the annual energy use of the *proposed design* and *standard reference design* based on estimates of energy use.

ENERGY COST. The total estimated annual cost for purchased energy for the building functions regulated by this code, including applicable demand charges.

ENERGY SIMULATION TOOL. An *approved* software program or calculation-based methodology that projects the annual energy use of a building.

ENTRANCE DOOR. Fenestration products used for ingress, egress and access in nonresidential buildings, including, but not limited to, exterior entrances that utilize latching hardware and automatic closers and contain over 50-percent glass specifically designed to withstand heavy use and possibly abuse.

EXTERIOR WALL. Walls including both above-grade walls and basement walls.

FENESTRATION. Skylights, roof windows, vertical windows (fixed or moveable), opaque doors, glazed doors, glazed block and combination opaque/glazed doors. Fenestration includes products with glass and nonglass glazing materials.

FENESTRATION PRODUCT, SITE-BUILT. A fenestration designed to be made up of field-glazed or field-assembled units using specific factory cut or otherwise factory-formed framing and glazing units. Examples of site-built fenestration include storefront systems, curtain walls, and atrium roof systems.

***F*-FACTOR.** The perimeter heat loss factor for slab-on-grade floors (Btu/h × ft × °F) [W/(m × K)].

HEATED SLAB. Slab-on-grade construction in which the heating elements, hydronic tubing, or hot air distribution system is in contact with, or placed within or under, the slab.

HIGH-EFFICACY LAMPS. Compact fluorescent lamps, T-8 or smaller diameter linear fluorescent lamps, or lamps with a minimum efficacy of:

1. 60 lumens per watt for lamps over 40 watts;
2. 50 lumens per watt for lamps over 15 watts to 40 watts; and
3. 40 lumens per watt for lamps 15 watts or less.

INFILTRATION. The uncontrolled inward air leakage into a building caused by the pressure effects of wind or the effect of differences in the indoor and outdoor air density or both.

INSULATING SHEATHING. An insulating board with a core material having a minimum *R*-value of R-2.

LABELED. Equipment, materials or products to which have been affixed a label, seal, symbol or other identifying mark of a nationally recognized testing laboratory, inspection agency or other organization concerned with product evaluation that maintains periodic inspection of the production of the above-labeled items and whose labeling indicates either that the equipment, material or product meets identified standards or has been tested and found suitable for a specified purpose.

LISTED. Equipment, materials, products or services included in a list published by an organization acceptable to the *code official* and concerned with evaluation of products or services that maintains periodic inspection of production of *listed* equipment or materials or periodic evaluation of services and whose listing states either that the equipment, material, product or service meets identified standards or has been tested and found suitable for a specified purpose.

LOW-VOLTAGE LIGHTING. Lighting equipment powered through a transformer such as a cable conductor, a rail conductor and track lighting.

MANUAL. Capable of being operated by personal intervention (see "Automatic").

PROPOSED DESIGN. A description of the proposed building used to estimate annual energy use for determining compliance based on total building performance.

READILY ACCESSIBLE. Capable of being reached quickly for operation, renewal or inspection without requiring those to whom ready access is requisite to climb over or remove obstacles or to resort to portable ladders or access equipment (see "*Accessible*").

REPAIR. The reconstruction or renewal of any part of an existing building.

RESIDENTIAL BUILDING. For this code, includes detached one- and two-family dwellings and multiple single-family dwellings (townhouses) as well as Group R-2, R-3 and R-4 buildings three stories or less in height above grade plane.

ROOF ASSEMBLY. A system designed to provide weather protection and resistance to design loads. The system consists of a roof covering and roof deck or a single component serving as both the roof covering and the roof deck. A roof assembly includes the roof covering, underlayment, roof deck, insulation, vapor retarder and interior finish.

***R*-VALUE (THERMAL RESISTANCE).** The inverse of the time rate of heat flow through a body from one of its bounding surfaces to the other surface for a unit temperature difference between the two surfaces, under steady state conditions, per unit area ($h \cdot \text{ft}^2 \cdot °\text{F/Btu}$) [$(\text{m}^2 \cdot \text{K})/\text{W}$].

SERVICE WATER HEATING. Supply of hot water for purposes other than comfort heating.

SKYLIGHT. Glass or other transparent or translucent glazing material installed at a slope of less than 60 degrees (1.05 rad) from horizontal. Glazing material in skylights, including unit skylights, solariums, sunrooms, roofs and sloped walls is included in this definition.

SOLAR HEAT GAIN COEFFICIENT (SHGC). The ratio of the solar heat gain entering the space through the fenestration assembly to the incident solar radiation. Solar heat gain includes directly transmitted solar heat and absorbed solar radiation which is then reradiated, conducted or convected into the space.

STANDARD REFERENCE DESIGN. A version of the *proposed design* that meets the minimum requirements of this code and is used to determine the maximum annual energy use requirement for compliance based on total building performance.

SUNROOM. A one-story structure attached to a dwelling with a glazing area in excess of 40 percent of the gross area of the structure's exterior walls and roof.

THERMAL ISOLATION. Physical and space conditioning separation from *conditioned space(s)*. The *conditioned space*(s) shall be controlled as separate zones for heating and cooling or conditioned by separate equipment.

THERMOSTAT. An automatic control device used to maintain temperature at a fixed or adjustable set point.

U-**FACTOR (THERMAL TRANSMITTANCE).** The coefficient of heat transmission (air to air) through a building component or assembly, equal to the time rate of heat flow per unit area and unit temperature difference between the warm side and cold side air films (Btu/h · ft^2 · °F) [W/(m^2 · K)].

VENTILATION. The natural or mechanical process of supplying conditioned or unconditioned air to, or removing such air from, any space.

VENTILATION AIR. That portion of supply air that comes from outside (outdoors) plus any recirculated air that has been treated to maintain the desired quality of air within a designated space.

VISIBLE TRANSMITTANCE [VT]. The ratio of visible light entering the space through the fenestration product assembly to the incident visible light, Visible Transmittance, includes the effects of glazing material and frame and is expressed as a number between 0 and 1.

WHOLE HOUSE MECHANICAL VENTILATION SYSTEM. An exhaust system, supply system, or combination thereof that is designed to mechanically exchange indoor air with outdoor air when operating continuously or through a programmed intermittent schedule to satisfy the whole house ventilation rates.

ZONE. A space or group of spaces within a building with heating or cooling requirements that are sufficiently similar so that desired conditions can be maintained throughout using a single controlling device.

CHAPTER 3 [RE]

GENERAL REQUIREMENTS

SECTION R301
CLIMATE ZONES

R301.1 General. Climate zones from Figure R301.1 or Table R301.1 shall be used in determining the applicable requirements from Chapter 4. Locations not in Table R301.1 (outside the United States) shall be assigned a climate zone based on Section R301.3.

R301.2 Warm humid counties. Warm humid counties are identified in Table R301.1 by an asterisk.

R301.3 International climate zones. The climate zone for any location outside the United States shall be determined by applying Table R301.3(1) and then Table R301.3(2).

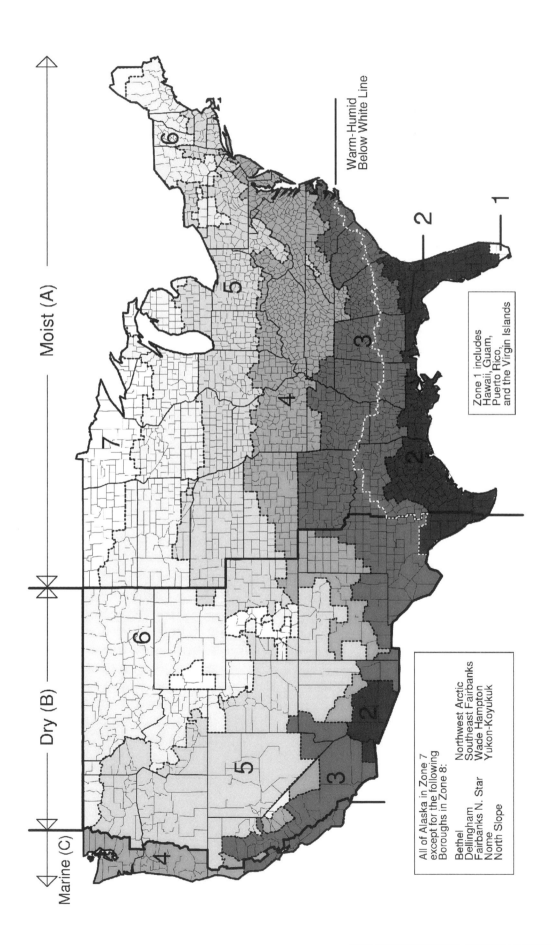

All of Alaska in Zone 7
except for the following
Boroughs in Zone 8:

Bethel	Northwest Arctic
Dellingham	Southeast Fairbanks
Fairbanks N. Star	Wade Hampton
Nome	Yukon-Koyukuk
North Slope	

Zone 1 includes
Hawaii, Guam,
Puerto Rico,
and the Virgin Islands

Warm-Humid
Below White Line

Moist (A)

Dry (B)

Marine (C)

**FIGURE R301.1
CLIMATE ZONES**

TABLE R301.1
CLIMATE ZONES, MOISTURE REGIMES, AND WARM-HUMID
DESIGNATIONS BY STATE, COUNTY AND TERRITORY

Key: A – Moist, B – Dry, C – Marine. Absence of moisture designation indicates moisture regime is irrelevant.
Asterisk (*) indicates a warm-humid location.

US STATES

ALABAMA

3A Autauga*
2A Baldwin*
3A Barbour*
3A Bibb
3A Blount
3A Bullock*
3A Butler*
3A Calhoun
3A Chambers
3A Cherokee
3A Chilton
3A Choctaw*
3A Clarke*
3A Clay
3A Cleburne
3A Coffee*
3A Colbert
3A Conecuh*
3A Coosa
3A Covington*
3A Crenshaw*
3A Cullman
3A Dale*
3A Dallas*
3A DeKalb
3A Elmore*
3A Escambia*
3A Etowah
3A Fayette
3A Franklin
3A Geneva*
3A Greene
3A Hale
3A Henry*
3A Houston*
3A Jackson
3A Jefferson
3A Lamar
3A Lauderdale
3A Lawrence

3A Lee
3A Limestone
3A Lowndes*
3A Macon*
3A Madison
3A Marengo*
3A Marion
3A Marshall
2A Mobile*
3A Monroe*
3A Montgomery*
3A Morgan
3A Perry*
3A Pickens
3A Pike*
3A Randolph
3A Russell*
3A Shelby
3A St. Clair
3A Sumter
3A Talladega
3A Tallapoosa
3A Tuscaloosa
3A Walker
3A Washington*
3A Wilcox*
3A Winston

ALASKA

7 Aleutians East
7 Aleutians West
7 Anchorage
8 Bethel
7 Bristol Bay
7 Denali
8 Dillingham
8 Fairbanks North Star
7 Haines
7 Juneau
7 Kenai Peninsula
7 Ketchikan Gateway

7 Kodiak Island
7 Lake and Peninsula
7 Matanuska-Susitna
8 Nome
8 North Slope
8 Northwest Arctic
7 Prince of Wales
   Outer Ketchikan
7 Sitka
7 Skagway-Hoonah-
   Angoon
8 Southeast Fairbanks
7 Valdez-Cordova
8 Wade Hampton
7 Wrangell-Petersburg
7 Yakutat
8 Yukon-Koyukuk

ARIZONA

5B Apache
3B Cochise
5B Coconino
4B Gila
3B Graham
3B Greenlee
2B La Paz
2B Maricopa
3B Mohave
5B Navajo
2B Pima
2B Pinal
3B Santa Cruz
4B Yavapai
2B Yuma

ARKANSAS

3A Arkansas
3A Ashley
4A Baxter
4A Benton
4A Boone
3A Bradley

3A Calhoun
4A Carroll
3A Chicot
3A Clark
3A Clay
3A Cleburne
3A Cleveland
3A Columbia*
3A Conway
3A Craighead
3A Crawford
3A Crittenden
3A Cross
3A Dallas
3A Desha
3A Drew
3A Faulkner
3A Franklin
4A Fulton
3A Garland
3A Grant
3A Greene
3A Hempstead*
3A Hot Spring
3A Howard
3A Independence
4A Izard
3A Jackson
3A Jefferson
3A Johnson
3A Lafayette*
3A Lawrence
3A Lee
3A Lincoln
3A Little River*
3A Logan
3A Lonoke
4A Madison
4A Marion
3A Miller*
3A Mississippi

3A Monroe
3A Montgomery
3A Nevada
4A Newton
3A Ouachita
3A Perry
3A Phillips
3A Pike
3A Poinsett
3A Polk
3A Pope
3A Prairie
3A Pulaski
3A Randolph
3A Saline
3A Scott
4A Searcy
3A Sebastian
3A Sevier*
3A Sharp
3A St. Francis
4A Stone
3A Union*
3A Van Buren
4A Washington
3A White
3A Woodruff
3A Yell

CALIFORNIA

3C Alameda
6B Alpine
4B Amador
3B Butte
4B Calaveras
3B Colusa
3B Contra Costa
4C Del Norte
4B El Dorado
3B Fresno
3B Glenn

(continued)

4C Humboldt
2B Imperial
4B Inyo
3B Kern
3B Kings
4B Lake
5B Lassen
3B Los Angeles
3B Madera
3C Marin
4B Mariposa
3C Mendocino
3B Merced
5B Modoc
6B Mono
3C Monterey
3C Napa
5B Nevada
3B Orange
3B Placer
5B Plumas
3B Riverside
3B Sacramento
3C San Benito
3B San Bernardino
3B San Diego
3C San Francisco
3B San Joaquin
3C San Luis Obispo
3C San Mateo
3C Santa Barbara
3C Santa Clara
3C Santa Cruz
3B Shasta
5B Sierra
5B Siskiyou
3B Solano
3C Sonoma
3B Stanislaus
3B Sutter
3B Tehama
4B Trinity
3B Tulare
4B Tuolumne
3C Ventura
3B Yolo

3B Yuba

COLORADO

5B Adams
6B Alamosa
5B Arapahoe
6B Archuleta
4B Baca
5B Bent
5B Boulder
6B Chaffee
5B Cheyenne
7 Clear Creek
6B Conejos
6B Costilla
5B Crowley
6B Custer
5B Delta
5B Denver
6B Dolores
5B Douglas
6B Eagle
5B Elbert
5B El Paso
5B Fremont
5B Garfield
5B Gilpin
7 Grand
7 Gunnison
7 Hinsdale
5B Huerfano
7 Jackson
5B Jefferson
5B Kiowa
5B Kit Carson
7 Lake
5B La Plata
5B Larimer
4B Las Animas
5B Lincoln
5B Logan
5B Mesa
7 Mineral
6B Moffat
5B Montezuma
5B Montrose

5B Morgan
4B Otero
6B Ouray
7 Park
5B Phillips
7 Pitkin
5B Prowers
5B Pueblo
6B Rio Blanco
7 Rio Grande
7 Routt
6B Saguache
7 San Juan
6B San Miguel
5B Sedgwick
7 Summit
5B Teller
5B Washington
5B Weld
5B Yuma

CONNECTICUT

5A (all)

DELAWARE

4A (all)

DISTRICT OF COLUMBIA

4A (all)

FLORIDA

2A Alachua*
2A Baker*
2A Bay*
2A Bradford*
2A Brevard*
1A Broward*
2A Calhoun*
2A Charlotte*
2A Citrus*
2A Clay*
2A Collier*
2A Columbia*
2A DeSoto*
2A Dixie*
2A Duval*

2A Escambia*
2A Flagler*
2A Franklin*
2A Gadsden*
2A Gilchrist*
2A Glades*
2A Gulf*
2A Hamilton*
2A Hardee*
2A Hendry*
2A Hernando*
2A Highlands*
2A Hillsborough*
2A Holmes*
2A Indian River*
2A Jackson*
2A Jefferson*
2A Lafayette*
2A Lake*
2A Lee*
2A Leon*
2A Levy*
2A Liberty*
2A Madison*
2A Manatee*
2A Marion*
2A Martin*
1A Miami-Dade*
1A Monroe*
2A Nassau*
2A Okaloosa*
2A Okeechobee*
2A Orange*
2A Osceola*
2A Palm Beach*
2A Pasco*
2A Pinellas*
2A Polk*
2A Putnam*
2A Santa Rosa*
2A Sarasota*
2A Seminole*
2A St. Johns*
2A St. Lucie*
2A Sumter*
2A Suwannee*

2A Taylor*
2A Union*
2A Volusia*
2A Wakulla*
2A Walton*
2A Washington*

GEORGIA

2A Appling*
2A Atkinson*
2A Bacon*
2A Baker*
3A Baldwin
4A Banks
3A Barrow
3A Bartow
3A Ben Hill*
2A Berrien*
3A Bibb
3A Bleckley*
2A Brantley*
2A Brooks*
2A Bryan*
3A Bulloch*
3A Burke
3A Butts
3A Calhoun*
2A Camden*
3A Candler*
3A Carroll
4A Catoosa
2A Charlton*
2A Chatham*
3A Chattahoochee*
4A Chattooga
3A Cherokee
3A Clarke
3A Clay*
3A Clayton
2A Clinch*
3A Cobb
3A Coffee*
2A Colquitt*
3A Columbia
2A Cook*
3A Coweta

(continued)

TABLE R301.1—continued
CLIMATE ZONES, MOISTURE REGIMES, AND WARM-HUMID
DESIGNATIONS BY STATE, COUNTY AND TERRITORY

3A Crawford	2A Lanier*	3A Taylor*	5B Cassia	4A Crawford
3A Crisp*	3A Laurens*	3A Telfair*	6B Clark	5A Cumberland
4A Dade	3A Lee*	3A Terrell*	5B Clearwater	5A DeKalb
4A Dawson	2A Liberty*	2A Thomas*	6B Custer	5A De Witt
2A Decatur*	3A Lincoln	3A Tift*	5B Elmore	5A Douglas
3A DeKalb	2A Long*	2A Toombs*	6B Franklin	5A DuPage
3A Dodge*	2A Lowndes*	4A Towns	6B Fremont	5A Edgar
3A Dooly*	4A Lumpkin	3A Treutlen*	5B Gem	4A Edwards
3A Dougherty*	3A Macon*	3A Troup	5B Gooding	4A Effingham
3A Douglas	3A Madison	3A Turner*	5B Idaho	4A Fayette
3A Early*	3A Marion*	3A Twiggs*	6B Jefferson	5A Ford
2A Echols*	3A McDuffie	4A Union	5B Jerome	4A Franklin
2A Effingham*	2A McIntosh*	3A Upson	5B Kootenai	5A Fulton
3A Elbert	3A Meriwether	4A Walker	5B Latah	4A Gallatin
3A Emanuel*	2A Miller*	3A Walton	6B Lemhi	5A Greene
2A Evans*	2A Mitchell*	2A Ware*	5B Lewis	5A Grundy
4A Fannin	3A Monroe	3A Warren	5B Lincoln	4A Hamilton
3A Fayette	3A Montgomery*	3A Washington	6B Madison	5A Hancock
4A Floyd	3A Morgan	2A Wayne*	5B Minidoka	4A Hardin
3A Forsyth	4A Murray	3A Webster*	5B Nez Perce	5A Henderson
4A Franklin	3A Muscogee	3A Wheeler*	6B Oneida	5A Henry
3A Fulton	3A Newton	4A White	5B Owyhee	5A Iroquois
4A Gilmer	3A Oconee	4A Whitfield	5B Payette	4A Jackson
3A Glascock	3A Oglethorpe	3A Wilcox*	5B Power	4A Jasper
2A Glynn*	3A Paulding	3A Wilkes	5B Shoshone	4A Jefferson
4A Gordon	3A Peach*	3A Wilkinson	6B Teton	5A Jersey
2A Grady*	4A Pickens	3A Worth*	5B Twin Falls	5A Jo Daviess
3A Greene	2A Pierce*		6B Valley	4A Johnson
3A Gwinnett	3A Pike	**HAWAII**	5B Washington	5A Kane
4A Habersham	3A Polk	1A (all)*		5A Kankakee
4A Hall	3A Pulaski*		**ILLINOIS**	5A Kendall
3A Hancock	3A Putnam	**IDAHO**	5A Adams	5A Knox
3A Haralson	3A Quitman*	5B Ada	4A Alexander	5A Lake
3A Harris	4A Rabun	6B Adams	4A Bond	5A La Salle
3A Hart	3A Randolph*	6B Bannock	5A Boone	4A Lawrence
3A Heard	3A Richmond	6B Bear Lake	5A Brown	5A Lee
3A Henry	3A Rockdale	5B Benewah	5A Bureau	5A Livingston
3A Houston*	3A Schley*	6B Bingham	5A Calhoun	5A Logan
3A Irwin*	3A Screven*	6B Blaine	5A Carroll	5A Macon
3A Jackson	2A Seminole*	6B Boise	5A Cass	4A Macoupin
3A Jasper	3A Spalding	6B Bonner	5A Champaign	4A Madison
2A Jeff Davis*	4A Stephens	6B Bonneville	4A Christian	4A Marion
3A Jefferson	3A Stewart*	6B Boundary	5A Clark	5A Marshall
3A Jenkins*	3A Sumter*	6B Butte	4A Clay	5A Mason
3A Johnson*	3A Talbot	6B Camas	4A Clinton	4A Massac
3A Jones	3A Taliaferro	5B Canyon	5A Coles	5A McDonough
3A Lamar	2A Tattnall*	6B Caribou	5A Cook	5A McHenry

(continued)

TABLE R301.1—continued
CLIMATE ZONES, MOISTURE REGIMES, AND WARM-HUMID
DESIGNATIONS BY STATE, COUNTY AND TERRITORY

5A McLean	5A Boone	5A Miami	5A Appanoose	5A Jasper
5A Menard	4A Brown	4A Monroe	5A Audubon	5A Jefferson
5A Mercer	5A Carroll	5A Montgomery	5A Benton	5A Johnson
4A Monroe	5A Cass	5A Morgan	6A Black Hawk	5A Jones
4A Montgomery	4A Clark	5A Newton	5A Boone	5A Keokuk
5A Morgan	5A Clay	5A Noble	6A Bremer	6A Kossuth
5A Moultrie	5A Clinton	4A Ohio	6A Buchanan	5A Lee
5A Ogle	4A Crawford	4A Orange	6A Buena Vista	5A Linn
5A Peoria	4A Daviess	5A Owen	6A Butler	5A Louisa
4A Perry	4A Dearborn	5A Parke	6A Calhoun	5A Lucas
5A Piatt	5A Decatur	4A Perry	5A Carroll	6A Lyon
5A Pike	5A De Kalb	4A Pike	5A Cass	5A Madison
4A Pope	5A Delaware	5A Porter	5A Cedar	5A Mahaska
4A Pulaski	4A Dubois	4A Posey	6A Cerro Gordo	5A Marion
5A Putnam	5A Elkhart	5A Pulaski	6A Cherokee	5A Marshall
4A Randolph	5A Fayette	5A Putnam	6A Chickasaw	5A Mills
4A Richland	4A Floyd	5A Randolph	5A Clarke	6A Mitchell
5A Rock Island	5A Fountain	4A Ripley	6A Clay	5A Monona
4A Saline	5A Franklin	5A Rush	6A Clayton	5A Monroe
5A Sangamon	5A Fulton	4A Scott	5A Clinton	5A Montgomery
5A Schuyler	4A Gibson	5A Shelby	5A Crawford	5A Muscatine
5A Scott	5A Grant	4A Spencer	5A Dallas	6A O'Brien
4A Shelby	4A Greene	5A Starke	5A Davis	6A Osceola
5A Stark	5A Hamilton	5A Steuben	5A Decatur	5A Page
4A St. Clair	5A Hancock	5A St. Joseph	6A Delaware	6A Palo Alto
5A Stephenson	4A Harrison	4A Sullivan	5A Des Moines	6A Plymouth
5A Tazewell	5A Hendricks	4A Switzerland	6A Dickinson	6A Pocahontas
4A Union	5A Henry	5A Tippecanoe	5A Dubuque	5A Polk
5A Vermilion	5A Howard	5A Tipton	6A Emmet	5A Pottawattamie
4A Wabash	5A Huntington	5A Union	6A Fayette	5A Poweshiek
5A Warren	4A Jackson	4A Vanderburgh	6A Floyd	5A Ringgold
4A Washington	5A Jasper	5A Vermillion	6A Franklin	6A Sac
4A Wayne	5A Jay	5A Vigo	5A Fremont	5A Scott
4A White	4A Jefferson	5A Wabash	5A Greene	5A Shelby
5A Whiteside	4A Jennings	5A Warren	6A Grundy	6A Sioux
5A Will	5A Johnson	4A Warrick	5A Guthrie	5A Story
4A Williamson	4A Knox	4A Washington	6A Hamilton	5A Tama
5A Winnebago	5A Kosciusko	5A Wayne	6A Hancock	5A Taylor
5A Woodford	5A Lagrange	5A Wells	6A Hardin	5A Union
	5A Lake	5A White	5A Harrison	5A Van Buren
INDIANA	5A La Porte	5A Whitley	5A Henry	5A Wapello
	4A Lawrence		6A Howard	5A Warren
5A Adams	5A Madison	**IOWA**	6A Humboldt	5A Washington
5A Allen	5A Marion		6A Ida	5A Wayne
5A Bartholomew	5A Marshall	5A Adair	5A Iowa	6A Webster
5A Benton	4A Martin	5A Adams	5A Jackson	6A Winnebago
5A Blackford		6A Allamakee		

(continued)

TABLE R301.1—continued
CLIMATE ZONES, MOISTURE REGIMES, AND WARM-HUMID
DESIGNATIONS BY STATE, COUNTY AND TERRITORY

6A Winneshiek
5A Woodbury
6A Worth
6A Wright

KANSAS

4A Allen
4A Anderson
4A Atchison
4A Barber
4A Barton
4A Bourbon
4A Brown
4A Butler
4A Chase
4A Chautauqua
4A Cherokee
5A Cheyenne
4A Clark
4A Clay
5A Cloud
4A Coffey
4A Comanche
4A Cowley
4A Crawford
5A Decatur
4A Dickinson
4A Doniphan
4A Douglas
4A Edwards
4A Elk
5A Ellis
4A Ellsworth
4A Finney
4A Ford
4A Franklin
4A Geary
5A Gove
5A Graham
4A Grant
4A Gray
5A Greeley
4A Greenwood
5A Hamilton
4A Harper
4A Harvey

4A Haskell
4A Hodgeman
4A Jackson
4A Jefferson
5A Jewell
4A Johnson
4A Kearny
4A Kingman
4A Kiowa
4A Labette
5A Lane
4A Leavenworth
4A Lincoln
4A Linn
5A Logan
4A Lyon
4A Marion
4A Marshall
4A McPherson
4A Meade
4A Miami
5A Mitchell
4A Montgomery
4A Morris
4A Morton
4A Nemaha
4A Neosho
5A Ness
5A Norton
4A Osage
5A Osborne
4A Ottawa
4A Pawnee
5A Phillips
4A Pottawatomie
4A Pratt
5A Rawlins
4A Reno
5A Republic
4A Rice
4A Riley
5A Rooks
4A Rush
4A Russell
4A Saline
5A Scott

4A Sedgwick
4A Seward
4A Shawnee
5A Sheridan
5A Sherman
5A Smith
4A Stafford
4A Stanton
4A Stevens
4A Sumner
5A Thomas
5A Trego
4A Wabaunsee
5A Wallace
4A Washington
5A Wichita
4A Wilson
4A Woodson
4A Wyandotte

KENTUCKY

4A (all)

LOUISIANA

2A Acadia*
2A Allen*
2A Ascension*
2A Assumption*
2A Avoyelles*
2A Beauregard*
3A Bienville*
3A Bossier*
3A Caddo*
2A Calcasieu*
3A Caldwell*
2A Cameron*
3A Catahoula*
3A Claiborne*
3A Concordia*
3A De Soto*
2A East Baton Rouge*
3A East Carroll
2A East Feliciana*
2A Evangeline*
3A Franklin*
3A Grant*
2A Iberia*

2A Iberville*
3A Jackson*
2A Jefferson*
2A Jefferson Davis*
2A Lafayette*
2A Lafourche*
3A La Salle*
3A Lincoln*
2A Livingston*
3A Madison*
3A Morehouse
3A Natchitoches*
2A Orleans*
3A Ouachita*
2A Plaquemines*
2A Pointe Coupee*
2A Rapides*
3A Red River*
3A Richland*
3A Sabine*
2A St. Bernard*
2A St. Charles*
2A St. Helena*
2A St. James*
2A St. John the
Baptist*
2A St. Landry*
2A St. Martin*
2A St. Mary*
2A St. Tammany*
2A Tangipahoa*
3A Tensas*
2A Terrebonne*
3A Union*
2A Vermilion*
3A Vernon*
2A Washington*
3A Webster*
2A West Baton
Rouge*
3A West Carroll
2A West Feliciana*
3A Winn*

MAINE

6A Androscoggin
7 Aroostook

6A Cumberland
6A Franklin
6A Hancock
6A Kennebec
6A Knox
6A Lincoln
6A Oxford
6A Penobscot
6A Piscataquis
6A Sagadahoc
6A Somerset
6A Waldo
6A Washington
6A York

MARYLAND

4A Allegany
4A Anne Arundel
4A Baltimore
4A Baltimore (city)
4A Calvert
4A Caroline
4A Carroll
4A Cecil
4A Charles
4A Dorchester
4A Frederick
5A Garrett
4A Harford
4A Howard
4A Kent
4A Montgomery
4A Prince George's
4A Queen Anne's
4A Somerset
4A St. Mary's
4A Talbot
4A Washington
4A Wicomico
4A Worcester

MASSACHSETTS

5A (all)

MICHIGAN

6A Alcona
6A Alger

(continued)

TABLE R301.1—continued
CLIMATE ZONES, MOISTURE REGIMES, AND WARM-HUMID
DESIGNATIONS BY STATE, COUNTY AND TERRITORY

5A Allegan
6A Alpena
6A Antrim
6A Arenac
7 Baraga
5A Barry
5A Bay
6A Benzie
5A Berrien
5A Branch
5A Calhoun
5A Cass
6A Charlevoix
6A Cheboygan
7 Chippewa
6A Clare
5A Clinton
6A Crawford
6A Delta
6A Dickinson
5A Eaton
6A Emmet
5A Genesee
6A Gladwin
7 Gogebic
6A Grand Traverse
5A Gratiot
5A Hillsdale
7 Houghton
6A Huron
5A Ingham
5A Ionia
6A Iosco
7 Iron
6A Isabella
5A Jackson
5A Kalamazoo
6A Kalkaska
5A Kent
7 Keweenaw
6A Lake
5A Lapeer
6A Leelanau
5A Lenawee
5A Livingston
7 Luce

7 Mackinac
5A Macomb
6A Manistee
6A Marquette
6A Mason
6A Mecosta
6A Menominee
5A Midland
6A Missaukee
5A Monroe
5A Montcalm
6A Montmorency
5A Muskegon
6A Newaygo
5A Oakland
6A Oceana
6A Ogemaw
7 Ontonagon
6A Osceola
6A Oscoda
6A Otsego
5A Ottawa
6A Presque Isle
6A Roscommon
5A Saginaw
6A Sanilac
7 Schoolcraft
5A Shiawassee
5A St. Clair
5A St. Joseph
5A Tuscola
5A Van Buren
5A Washtenaw
5A Wayne
6A Wexford

MINNESOTA

7 Aitkin
6A Anoka
7 Becker
7 Beltrami
6A Benton
6A Big Stone
6A Blue Earth
6A Brown
7 Carlton

6A Carver
7 Cass
6A Chippewa
6A Chisago
7 Clay
7 Clearwater
7 Cook
6A Cottonwood
7 Crow Wing
6A Dakota
6A Dodge
6A Douglas
6A Faribault
6A Fillmore
6A Freeborn
6A Goodhue
7 Grant
6A Hennepin
6A Houston
7 Hubbard
6A Isanti
7 Itasca
6A Jackson
7 Kanabec
6A Kandiyohi
7 Kittson
7 Koochiching
6A Lac qui Parle
7 Lake
7 Lake of the Woods
6A Le Sueur
6A Lincoln
6A Lyon
7 Mahnomen
7 Marshall
6A Martin
6A McLeod
6A Meeker
7 Mille Lacs
6A Morrison
6A Mower
6A Murray
6A Nicollet
6A Nobles
7 Norman
6A Olmsted

7 Otter Tail
7 Pennington
7 Pine
6A Pipestone
7 Polk
6A Pope
6A Ramsey
7 Red Lake
6A Redwood
6A Renville
6A Rice
6A Rock
7 Roseau
6A Scott
6A Sherburne
6A Sibley
6A Stearns
6A Steele
6A Stevens
7 St. Louis
6A Swift
6A Todd
6A Traverse
6A Wabasha
7 Wadena
6A Waseca
6A Washington
6A Watonwan
7 Wilkin
6A Winona
6A Wright
6A Yellow
　Medicine

MISSISSIPPI

3A Adams*
3A Alcorn
3A Amite*
3A Attala
3A Benton
3A Bolivar
3A Calhoun
3A Carroll
3A Chickasaw
3A Choctaw
3A Claiborne*

3A Clarke
3A Clay
3A Coahoma
3A Copiah*
3A Covington*
3A DeSoto
3A Forrest*
3A Franklin*
3A George*
3A Greene*
3A Grenada
2A Hancock*
2A Harrison*
3A Hinds*
3A Holmes
3A Humphreys
3A Issaquena
3A Itawamba
2A Jackson*
3A Jasper
3A Jefferson*
3A Jefferson Davis*
3A Jones*
3A Kemper
3A Lafayette
3A Lamar*
3A Lauderdale
3A Lawrence*
3A Leake
3A Lee
3A Leflore
3A Lincoln*
3A Lowndes
3A Madison
3A Marion*
3A Marshall
3A Monroe
3A Montgomery
3A Neshoba
3A Newton
3A Noxubee
3A Oktibbeha
3A Panola
2A Pearl River*
3A Perry*
3A Pike*

(continued)

2012 INTERNATIONAL ENERGY CONSERVATION CODE®

TABLE R301.1—continued
CLIMATE ZONES, MOISTURE REGIMES, AND WARM-HUMID
DESIGNATIONS BY STATE, COUNTY AND TERRITORY

3A Pontotoc
3A Prentiss
3A Quitman
3A Rankin*
3A Scott
3A Sharkey
3A Simpson*
3A Smith*
2A Stone*
3A Sunflower
3A Tallahatchie
3A Tate
3A Tippah
3A Tishomingo
3A Tunica
3A Union
3A Walthall*
3A Warren*
3A Washington
3A Wayne*
3A Webster
3A Wilkinson*
3A Winston
3A Yalobusha
3A Yazoo

MISSOURI

5A Adair
5A Andrew
5A Atchison
4A Audrain
4A Barry
4A Barton
4A Bates
4A Benton
4A Bollinger
4A Boone
5A Buchanan
4A Butler
5A Caldwell
4A Callaway
4A Camden
4A Cape Girardeau
4A Carroll
4A Carter
4A Cass
4A Cedar

5A Chariton
4A Christian
5A Clark
4A Clay
5A Clinton
4A Cole
4A Cooper
4A Crawford
4A Dade
4A Dallas
5A Daviess
5A DeKalb
4A Dent
4A Douglas
4A Dunklin
4A Franklin
4A Gasconade
5A Gentry
4A Greene
5A Grundy
5A Harrison
4A Henry
4A Hickory
5A Holt
4A Howard
4A Howell
4A Iron
4A Jackson
4A Jasper
4A Jefferson
4A Johnson
5A Knox
4A Laclede
4A Lafayette
4A Lawrence
5A Lewis
4A Lincoln
5A Linn
5A Livingston
5A Macon
4A Madison
4A Maries
5A Marion
4A McDonald
5A Mercer
4A Miller

4A Mississippi
4A Moniteau
4A Monroe
4A Montgomery
4A Morgan
4A New Madrid
4A Newton
5A Nodaway
4A Oregon
4A Osage
4A Ozark
4A Pemiscot
4A Perry
4A Pettis
4A Phelps
5A Pike
4A Platte
4A Polk
4A Pulaski
5A Putnam
5A Ralls
4A Randolph
4A Ray
4A Reynolds
4A Ripley
4A Saline
5A Schuyler
5A Scotland
4A Scott
4A Shannon
5A Shelby
4A St. Charles
4A St. Clair
4A Ste. Genevieve
4A St. Francois
4A St. Louis
4A St. Louis (city)
4A Stoddard
4A Stone
5A Sullivan
4A Taney
4A Texas
4A Vernon
4A Warren
4A Washington
4A Wayne

4A Webster
5A Worth
4A Wright

MONTANA

6B (all)

NEBRASKA

5A (all)

NEVADA

5B Carson City (city)
5B Churchill
3B Clark
5B Douglas
5B Elko
5B Esmeralda
5B Eureka
5B Humboldt
5B Lander
5B Lincoln
5B Lyon
5B Mineral
5B Nye
5B Pershing
5B Storey
5B Washoe
5B White Pine

NEW HAMPSHIRE

6A Belknap
6A Carroll
5A Cheshire
6A Coos
6A Grafton
5A Hillsborough
6A Merrimack
5A Rockingham
5A Strafford
6A Sullivan

NEW JERSEY

4A Atlantic
5A Bergen
4A Burlington
4A Camden
4A Cape May

4A Cumberland
4A Essex
4A Gloucester
4A Hudson
5A Hunterdon
5A Mercer
4A Middlesex
4A Monmouth
5A Morris
4A Ocean
5A Passaic
4A Salem
5A Somerset
5A Sussex
4A Union
5A Warren

NEW MEXICO

4B Bernalillo
5B Catron
3B Chaves
4B Cibola
5B Colfax
4B Curry
4B DeBaca
3B Dona Ana
3B Eddy
4B Grant
4B Guadalupe
5B Harding
3B Hidalgo
3B Lea
4B Lincoln
5B Los Alamos
3B Luna
5B McKinley
5B Mora
3B Otero
4B Quay
5B Rio Arriba
4B Roosevelt
5B Sandoval
5B San Juan
5B San Miguel
5B Santa Fe
4B Sierra
4B Socorro

(continued)

TABLE R301.1—continued
CLIMATE ZONES, MOISTURE REGIMES, AND WARM-HUMID
DESIGNATIONS BY STATE, COUNTY AND TERRITORY

5B Taos	4A Queens	4A Clay	4A Orange	7 Divide
5B Torrance	5A Rensselaer	4A Cleveland	3A Pamlico	6A Dunn
4B Union	4A Richmond	3A Columbus*	3A Pasquotank	7 Eddy
4B Valencia	5A Rockland	3A Craven	3A Pender*	6A Emmons
	5A Saratoga	3A Cumberland	3A Perquimans	7 Foster
NEW YORK	5A Schenectady	3A Currituck	4A Person	6A Golden Valley
	6A Schoharie	3A Dare	3A Pitt	7 Grand Forks
5A Albany	6A Schuyler	3A Davidson	4A Polk	6A Grant
6A Allegany	5A Seneca	4A Davie	3A Randolph	7 Griggs
4A Bronx	6A Steuben	3A Duplin	3A Richmond	6A Hettinger
6A Broome	6A St. Lawrence	4A Durham	3A Robeson	7 Kidder
6A Cattaraugus	4A Suffolk	3A Edgecombe	4A Rockingham	6A LaMoure
5A Cayuga	6A Sullivan	4A Forsyth	3A Rowan	6A Logan
5A Chautauqua	5A Tioga	4A Franklin	4A Rutherford	7 McHenry
5A Chemung	6A Tompkins	3A Gaston	3A Sampson	6A McIntosh
6A Chenango	6A Ulster	4A Gates	3A Scotland	6A McKenzie
6A Clinton	6A Warren	4A Graham	3A Stanly	7 McLean
5A Columbia	5A Washington	4A Granville	4A Stokes	6A Mercer
5A Cortland	5A Wayne	3A Greene	4A Surry	6A Morton
6A Delaware	4A Westchester	4A Guilford	4A Swain	7 Mountrail
5A Dutchess	6A Wyoming	4A Halifax	4A Transylvania	7 Nelson
5A Erie	5A Yates	4A Harnett	3A Tyrrell	6A Oliver
6A Essex		4A Haywood	3A Union	7 Pembina
6A Franklin	**NORTH CAROLINA**	4A Henderson	4A Vance	7 Pierce
6A Fulton		4A Hertford	4A Wake	7 Ramsey
5A Genesee	4A Alamance	3A Hoke	4A Warren	6A Ransom
5A Greene	4A Alexander	3A Hyde	3A Washington	7 Renville
6A Hamilton	5A Alleghany	4A Iredell	5A Watauga	6A Richland
6A Herkimer	3A Anson	4A Jackson	3A Wayne	7 Rolette
6A Jefferson	5A Ashe	3A Johnston	4A Wilkes	6A Sargent
4A Kings	5A Avery	3A Jones	3A Wilson	7 Sheridan
6A Lewis	3A Beaufort	4A Lee	4A Yadkin	6A Sioux
5A Livingston	4A Bertie	3A Lenoir	5A Yancey	6A Slope
6A Madison	3A Bladen	4A Lincoln		6A Stark
5A Monroe	3A Brunswick*	4A Macon	**NORTH DAKOTA**	7 Steele
6A Montgomery	4A Buncombe	4A Madison		7 Stutsman
4A Nassau	4A Burke	3A Martin	6A Adams	7 Towner
4A New York	3A Cabarrus	4A McDowell	7 Barnes	7 Traill
5A Niagara	4A Caldwell	3A Mecklenburg	7 Benson	7 Walsh
6A Oneida	3A Camden	5A Mitchell	6A Billings	7 Walsh
5A Onondaga	3A Carteret*	3A Montgomery	7 Bottineau	7 Ward
5A Ontario	4A Caswell	3A Moore	6A Bowman	7 Wells
5A Orange	4A Catawba	4A Nash	7 Burke	7 Williams
5A Orleans	4A Chatham	3A New Hanover*	6A Burleigh	
5A Oswego	4A Cherokee	4A Northampton	7 Cass	**OHIO**
6A Otsego	3A Chowan	3A Onslow*	7 Cavalier	4A Adams
5A Putnam			6A Dickey	5A Allen

(continued)

5A Ashland	5A Mahoning	3A Bryan	3A Okfuskee	4C Linn
5A Ashtabula	5A Marion	3A Caddo	3A Oklahoma	5B Malheur
5A Athens	5A Medina	3A Canadian	3A Okmulgee	4C Marion
5A Auglaize	5A Meigs	3A Carter	3A Osage	5B Morrow
5A Belmont	5A Mercer	3A Cherokee	3A Ottawa	4C Multnomah
4A Brown	5A Miami	3A Choctaw	3A Pawnee	4C Polk
5A Butler	5A Monroe	4B Cimarron	3A Payne	5B Sherman
5A Carroll	5A Montgomery	3A Cleveland	3A Pittsburg	4C Tillamook
5A Champaign	5A Morgan	3A Coal	3A Pontotoc	5B Umatilla
5A Clark	5A Morrow	3A Comanche	3A Pottawatomie	5B Union
4A Clermont	5A Muskingum	3A Cotton	3A Pushmataha	5B Wallowa
5A Clinton	5A Noble	3A Craig	3A Roger Mills	5B Wasco
5A Columbiana	5A Ottawa	3A Creek	3A Rogers	4C Washington
5A Coshocton	5A Paulding	3A Custer	3A Seminole	5B Wheeler
5A Crawford	5A Perry	3A Delaware	3A Sequoyah	4C Yamhill
5A Cuyahoga	5A Pickaway	3A Dewey	3A Stephens	
5A Darke	4A Pike	3A Ellis	4B Texas	**PENNSYLVANIA**
5A Defiance	5A Portage	3A Garfield	3A Tillman	5A Adams
5A Delaware	5A Preble	3A Garvin	3A Tulsa	5A Allegheny
5A Erie	5A Putnam	3A Grady	3A Wagoner	5A Armstrong
5A Fairfield	5A Richland	3A Grant	3A Washington	5A Beaver
5A Fayette	5A Ross	3A Greer	3A Washita	5A Bedford
5A Franklin	5A Sandusky	3A Harmon	3A Woods	5A Berks
5A Fulton	4A Scioto	3A Harper	3A Woodward	5A Blair
4A Gallia	5A Seneca	3A Haskell		5A Bradford
5A Geauga	5A Shelby	3A Hughes	**OREGON**	4A Bucks
5A Greene	5A Stark	3A Jackson	5B Baker	5A Butler
5A Guernsey	5A Summit	3A Jefferson	4C Benton	5A Cambria
4A Hamilton	5A Trumbull	3A Johnston	4C Clackamas	6A Cameron
5A Hancock	5A Tuscarawas	3A Kay	4C Clatsop	5A Carbon
5A Hardin	5A Union	3A Kingfisher	4C Columbia	5A Centre
5A Harrison	5A Van Wert	3A Kiowa	4C Coos	4A Chester
5A Henry	5A Vinton	3A Latimer	5B Crook	5A Clarion
5A Highland	5A Warren	3A Le Flore	4C Curry	6A Clearfield
5A Hocking	4A Washington	3A Lincoln	5B Deschutes	5A Clinton
5A Holmes	5A Wayne	3A Logan	4C Douglas	5A Columbia
5A Huron	5A Williams	3A Love	5B Gilliam	5A Crawford
5A Jackson	5A Wood	3A Major	5B Grant	5A Cumberland
5A Jefferson	5A Wyandot	3A Marshall	5B Harney	5A Dauphin
5A Knox		3A Mayes	5B Hood River	4A Delaware
5A Lake	**OKLAHOMA**	3A McClain	4C Jackson	6A Elk
4A Lawrence		3A McCurtain	5B Jefferson	5A Erie
5A Licking	3A Adair	3A McIntosh	4C Josephine	5A Fayette
5A Logan	3A Alfalfa	3A Murray	5B Klamath	5A Forest
5A Lorain	3A Atoka	3A Muskogee	5B Lake	5A Franklin
5A Lucas	4B Beaver	3A Noble	4C Lane	5A Fulton
5A Madison	3A Beckham	3A Nowata	4C Lincoln	5A Greene
	3A Blaine			

(continued)

TABLE R301.1—continued
CLIMATE ZONES, MOISTURE REGIMES, AND WARM-HUMID
DESIGNATIONS BY STATE, COUNTY AND TERRITORY

5A Huntingdon	3A Bamberg*	5A Bennett	6A Minnehaha	4A Gibson
5A Indiana	3A Barnwell*	5A Bon Homme	6A Moody	4A Giles
5A Jefferson	3A Beaufort*	6A Brookings	6A Pennington	4A Grainger
5A Juniata	3A Berkeley*	6A Brown	6A Perkins	4A Greene
5A Lackawanna	3A Calhoun	6A Brule	6A Potter	4A Grundy
5A Lancaster	3A Charleston*	6A Buffalo	6A Roberts	4A Hamblen
5A Lawrence	3A Cherokee	6A Butte	6A Sanborn	4A Hamilton
5A Lebanon	3A Chester	6A Campbell	6A Shannon	4A Hancock
5A Lehigh	3A Chesterfield	5A Charles Mix	6A Spink	3A Hardeman
5A Luzerne	3A Clarendon	6A Clark	6A Stanley	3A Hardin
5A Lycoming	3A Colleton*	5A Clay	6A Sully	4A Hawkins
6A McKean	3A Darlington	6A Codington	5A Todd	3A Haywood
5A Mercer	3A Dillon	6A Corson	5A Tripp	3A Henderson
5A Mifflin	3A Dorchester*	6A Custer	6A Turner	4A Henry
5A Monroe	3A Edgefield	6A Davison	5A Union	4A Hickman
4A Montgomery	3A Fairfield	6A Day	6A Walworth	4A Houston
5A Montour	3A Florence	6A Deuel	5A Yankton	4A Humphreys
5A Northampton	3A Georgetown*	6A Dewey	6A Ziebach	4A Jackson
5A Northumberland	3A Greenville	5A Douglas		4A Jefferson
5A Perry	3A Greenwood	6A Edmunds	**TENNESSEE**	4A Johnson
4A Philadelphia	3A Hampton*	6A Fall River	4A Anderson	4A Knox
5A Pike	3A Horry*	6A Faulk	4A Bedford	3A Lake
6A Potter	3A Jasper*	6A Grant	4A Benton	3A Lauderdale
5A Schuylkill	3A Kershaw	5A Gregory	4A Bledsoe	4A Lawrence
5A Snyder	3A Lancaster	6A Haakon	4A Blount	4A Lewis
5A Somerset	3A Laurens	6A Hamlin	4A Bradley	4A Lincoln
5A Sullivan	3A Lee	6A Hand	4A Campbell	4A Loudon
6A Susquehanna	3A Lexington	6A Hanson	4A Cannon	4A Macon
6A Tioga	3A Marion	6A Harding	4A Carroll	3A Madison
5A Union	3A Marlboro	6A Hughes	4A Carter	4A Marion
5A Venango	3A McCormick	5A Hutchinson	4A Cheatham	4A Marshall
5A Warren	3A Newberry	6A Hyde	3A Chester	4A Maury
5A Washington	3A Oconee	5A Jackson	4A Claiborne	4A McMinn
6A Wayne	3A Orangeburg	6A Jerauld	4A Clay	3A McNairy
5A Westmoreland	3A Pickens	6A Jones	4A Cocke	4A Meigs
5A Wyoming	3A Richland	6A Kingsbury	4A Coffee	4A Monroe
4A York	3A Saluda	6A Lake	3A Crockett	4A Montgomery
RHODE ISLAND	3A Spartanburg	6A Lawrence	4A Cumberland	4A Moore
	3A Sumter	6A Lincoln	4A Davidson	4A Morgan
5A (all)	3A Union	6A Lyman	4A Decatur	4A Obion
SOUTH	3A Williamsburg	6A Marshall	4A DeKalb	4A Overton
CAROLINA	3A York	6A McCook	4A Dickson	4A Perry
3A Abbeville	**SOUTH DAKOTA**	6A McPherson	3A Dyer	4A Pickett
3A Aiken		6A Meade	3A Fayette	4A Polk
3A Allendale*	6A Aurora	5A Mellette	4A Fentress	4A Putnam
3A Anderson	6A Beadle	6A Miner	4A Franklin	4A Rhea

(continued)

TABLE R301.1—continued
CLIMATE ZONES, MOISTURE REGIMES, AND WARM-HUMID
DESIGNATIONS BY STATE, COUNTY AND TERRITORY

4A Roane	3B Brewster	3B Ector	3B Howard	3B McCulloch
4A Robertson	4B Briscoe	2B Edwards*	3B Hudspeth	2A McLennan*
4A Rutherford	2A Brooks*	3A Ellis*	3A Hunt*	2A McMullen*
4A Scott	3A Brown*	3B El Paso	4B Hutchinson	2B Medina*
4A Sequatchie	2A Burleson*	3A Erath*	3B Irion	3B Menard
4A Sevier	3A Burnet*	2A Falls*	3A Jack	3B Midland
3A Shelby	2A Caldwell*	3A Fannin	2A Jackson*	2A Milam*
4A Smith	2A Calhoun*	2A Fayette*	2A Jasper*	3A Mills*
4A Stewart	3B Callahan	3B Fisher	3B Jeff Davis	3B Mitchell
4A Sullivan	2A Cameron*	4B Floyd	2A Jefferson*	3A Montague
4A Sumner	3A Camp*	3B Foard	2A Jim Hogg*	2A Montgomery*
3A Tipton	4B Carson	2A Fort Bend*	2A Jim Wells*	4B Moore
4A Trousdale	3A Cass*	3A Franklin*	3A Johnson*	3A Morris*
4A Unicoi	4B Castro	2A Freestone*	3B Jones	3B Motley
4A Union	2A Chambers*	2B Frio*	2A Karnes*	3A Nacogdoches*
4A Van Buren	2A Cherokee*	3B Gaines	3A Kaufman*	3A Navarro*
4A Warren	3B Childress	2A Galveston*	3A Kendall*	2A Newton*
4A Washington	3A Clay	3B Garza	2A Kenedy*	3B Nolan
4A Wayne	4B Cochran	3A Gillespie*	3B Kent	2A Nueces*
4A Weakley	3B Coke	3B Glasscock	3B Kerr	4B Ochiltree
4A White	3B Coleman	2A Goliad*	3B Kimble	4B Oldham
4A Williamson	3A Collin*	2A Gonzales*	3B King	2A Orange*
4A Wilson	3B Collingsworth	4B Gray	2B Kinney*	3A Palo Pinto*
	2A Colorado*	3A Grayson	2A Kleberg*	3A Panola*
TEXAS	2A Comal*	3A Gregg*	3B Knox	3A Parker*
	3A Comanche*	2A Grimes*	3A Lamar*	4B Parmer
2A Anderson*	3B Concho	2A Guadalupe*	4B Lamb	3B Pecos
3B Andrews	3A Cooke	4B Hale	3A Lampasas*	2A Polk*
2A Angelina*	2A Coryell*	3B Hall	2B La Salle*	4B Potter
2A Aransas*	3B Cottle	3A Hamilton*	2A Lavaca*	3B Presidio
3A Archer	3B Crane	4B Hansford	2A Lee*	3A Rains*
4B Armstrong	3B Crockett	3B Hardeman	2A Leon*	4B Randall
2A Atascosa*	3B Crosby	2A Hardin*	2A Liberty*	3B Reagan
2A Austin*	3B Culberson	2A Harris*	2A Limestone*	2B Real*
4B Bailey	4B Dallam	3A Harrison*	4B Lipscomb	3A Red River*
2B Bandera*	3A Dallas*	4B Hartley	2A Live Oak*	3B Reeves
2A Bastrop*	3B Dawson	3B Haskell	3A Llano*	2A Refugio*
3B Baylor	4B Deaf Smith	2A Hays*	3B Loving	4B Roberts
2A Bee*	3A Delta	3B Hemphill	3B Lubbock	2A Robertson*
2A Bell*	3A Denton*	3A Henderson*	3B Lynn	3A Rockwall*
2A Bexar*	2A DeWitt*	2A Hidalgo*	2A Madison*	3B Runnels
3A Blanco*	3B Dickens	2A Hill*	3A Marion*	3A Rusk*
3B Borden	2B Dimmit*	4B Hockley	3B Martin	3A Sabine*
2A Bosque*	4B Donley	3A Hood*	3B Mason	3A San Augustine*
3A Bowie*	2A Duval*	3A Hopkins*	2A Matagorda*	2A San Jacinto*
2A Brazoria*	3A Eastland	2A Houston*	2B Maverick*	2A San Patricio*
2A Brazos*				

(continued)

TABLE R301.1—continued
CLIMATE ZONES, MOISTURE REGIMES, AND WARM-HUMID
DESIGNATIONS BY STATE, COUNTY AND TERRITORY

3A San Saba*
3B Schleicher
3B Scurry
3B Shackelford
3A Shelby*
4B Sherman
3A Smith*
3A Somervell*
2A Starr*
3A Stephens
3B Sterling
3B Stonewall
3B Sutton
4B Swisher
3A Tarrant*
3B Taylor
3B Terrell
3B Terry
3B Throckmorton
3A Titus*
3B Tom Green
2A Travis*
2A Trinity*
2A Tyler*
3A Upshur*
3B Upton
2B Uvalde*
2B Val Verde*
3A Van Zandt*
2A Victoria*
2A Walker*
2A Waller*
3B Ward
2A Washington*
2B Webb*
2A Wharton*
3B Wheeler
3A Wichita
3B Wilbarger
2A Willacy*
2A Williamson*
2A Wilson*
3B Winkler
3A Wise
3A Wood*
4B Yoakum

3A Young
2B Zapata*
2B Zavala*

UTAH

5B Beaver
6B Box Elder
6B Cache
6B Carbon
6B Daggett
5B Davis
6B Duchesne
5B Emery
5B Garfield
5B Grand
5B Iron
5B Juab
5B Kane
5B Millard
6B Morgan
5B Piute
6B Rich
5B Salt Lake
5B San Juan
5B Sanpete
5B Sevier
6B Summit
5B Tooele
6B Uintah
5B Utah
6B Wasatch
3B Washington
5B Wayne
5B Weber

VERMONT

6A (all)

VIRGINIA

4A (all)

WASHINGTON

5B Adams
5B Asotin
5B Benton
5B Chelan
4C Clallam

4C Clark
5B Columbia
4C Cowlitz
5B Douglas
6B Ferry
5B Franklin
5B Garfield
5B Grant
4C Grays Harbor
4C Island
4C Jefferson
4C King
4C Kitsap
5B Kittitas
5B Klickitat
4C Lewis
5B Lincoln
4C Mason
6B Okanogan
4C Pacific
6B Pend Oreille
4C Pierce
4C San Juan
4C Skagit
5B Skamania
4C Snohomish
5B Spokane
6B Stevens
4C Thurston
4C Wahkiakum
5B Walla Walla
4C Whatcom
5B Whitman
5B Yakima

WEST VIRGINIA

5A Barbour
4A Berkeley
4A Boone
4A Braxton
5A Brooke
4A Cabell
4A Calhoun
4A Clay
5A Doddridge
5A Fayette

4A Gilmer
5A Grant
5A Greenbrier
5A Hampshire
5A Hancock
5A Hardy
5A Harrison
4A Jackson
4A Jefferson
4A Kanawha
5A Lewis
4A Lincoln
4A Logan
5A Marion
5A Marshall
4A Mason
4A McDowell
4A Mercer
5A Mineral
4A Mingo
5A Monongalia
4A Monroe
4A Morgan
5A Nicholas
5A Ohio
5A Pendleton
4A Pleasants
5A Pocahontas
5A Preston
4A Putnam
5A Raleigh
5A Randolph
4A Ritchie
4A Roane
5A Summers
5A Taylor
5A Tucker
4A Tyler
5A Upshur
4A Wayne
5A Webster
5A Wetzel
4A Wirt
4A Wood
4A Wyoming

WISCONSIN

6A Adams
7 Ashland
6A Barron
7 Bayfield
6A Brown
6A Buffalo
7 Burnett
6A Calumet
6A Chippewa
6A Clark
6A Columbia
6A Crawford
6A Dane
6A Dodge
6A Door
7 Douglas
6A Dunn
6A Eau Claire
7 Florence
6A Fond du Lac
7 Forest
6A Grant
6A Green
6A Green Lake
6A Iowa
7 Iron
6A Jackson
6A Jefferson
6A Juneau
6A Kenosha
6A Kewaunee
6A La Crosse
6A Lafayette
7 Langlade
7 Lincoln
6A Manitowoc
6A Marathon
6A Marinette
6A Marquette
6A Menominee
6A Milwaukee
6A Monroe
6A Oconto
7 Oneida
6A Outagamie

(continued)

TABLE R301.1—continued
CLIMATE ZONES, MOISTURE REGIMES, AND WARM-HUMID
DESIGNATIONS BY STATE, COUNTY AND TERRITORY

6A Ozaukee	7 Taylor	6B Big Horn	6B Sheridan	**NORTHERN**
6A Pepin	6A Trempealeau	6B Campbell	7 Sublette	**MARIANA**
6A Pierce	6A Vernon	6B Carbon	6B Sweetwater	**ISLANDS**
6A Polk	7 Vilas	6B Converse	7 Teton	1A (all)*
6A Portage	6A Walworth	6B Crook	6B Uinta	
7 Price	7 Washburn	6B Fremont	6B Washakie	**PUERTO RICO**
6A Racine	6A Washington	5B Goshen	6B Weston	1A (all)*
6A Richland	6A Waukesha	6B Hot Springs		
6A Rock	6A Waupaca	6B Johnson	**US TERRITORIES**	**VIRGIN ISLANDS**
6A Rusk	6A Waushara	6B Laramie		1A (all)*
6A Sauk	6A Winnebago	7 Lincoln	**AMERICAN**	
7 Sawyer	6A Wood	6B Natrona	**SAMOA**	
6A Shawano		6B Niobrara	1A (all)*	
6A Sheboygan	**WYOMING**	6B Park	**GUAM**	
6A St. Croix	6B Albany	5B Platte	1A (all)*	

TABLE R301.3(1)
INTERNATIONAL CLIMATE ZONE DEFINITIONS

MAJOR CLIMATE TYPE DEFINITIONS
Marine (C) Definition—Locations meeting all four criteria:

1. Mean temperature of coldest month between -3°C (27°F) and 18°C (65°F).

2. Warmest month mean < 22°C (72°F).

3. At least four months with mean temperatures over 10°C (50°F).

4. Dry season in summer. The month with the heaviest precipitation in the cold season has at least three times as much precipitation as the month with the least precipitation in the rest of the year. The cold season is October through March in the Northern Hemisphere and April through September in the Southern Hemisphere.

Dry (B) Definition—Locations meeting the following criteria:

Not marine and $P_{in} < 0.44 \times (TF - 19.5)$ [$P_{cm} < 2.0 \times (TC + 7)$ in SI units]
where:

P_{in} = Annual precipitation in inches (cm)

T = Annual mean temperature in °F (°C)

Moist (A) Definition—Locations that are not marine and not dry.

Warm-humid Definition—Moist (A) locations where either of the following wet-bulb temperature conditions shall occur during the warmest six consecutive months of the year:

1. 67°F (19.4°C) or higher for 3,000 or more hours; or

2. 73°F (22.8°C) or higher for 1,500 or more hours.

For SI: °C = [(°F)-32]/1.8, 1 inch = 2.54 cm.

TABLE R301.3(2)
INTERNATIONAL CLIMATE ZONE DEFINITIONS

ZONE NUMBER	THERMAL CRITERIA	
	IP Units	SI Units
1	9000 < CDD50°F	5000 < CDD10°C
2	6300 < CDD50°F ≤ 9000	3500 < CDD10°C ≤ 5000
3A and 3B	4500 < CDD50°F ≤ 6300 AND HDD65°F ≤ 5400	2500 < CDD10°C ≤ 3500 AND HDD18°C ≤ 3000
4A and 4B	CDD50°F ≤ 4500 AND HDD65°F ≤ 5400	CDD10°C ≤ 2500 AND HDD18°C ≤ 3000
3C	HDD65°F ≤ 3600	HDD18°C ≤ 2000
4C	3600 < HDD65°F ≤ 5400	2000 < HDD18°C ≤ 3000
5	5400 < HDD65°F ≤ 7200	3000 < HDD18°C ≤ 4000
6	7200 < HDD65°F ≤ 9000	4000 < HDD18°C ≤ 5000
7	9000 < HDD65°F ≤ 12600	5000 < HDD18°C ≤ 7000
8	12600 < HDD65°F	7000 < HDD18°C

For SI: °C = [(°F)-32]/1.8.

SECTION R302
DESIGN CONDITIONS

R302.1 Interior design conditions. The interior design temperatures used for heating and cooling load calculations shall be a maximum of 72°F (22°C) for heating and minimum of 75°F (24°C) for cooling.

SECTION R303
MATERIALS, SYSTEMS AND EQUIPMENT

R303.1 Identification. Materials, systems and equipment shall be identified in a manner that will allow a determination of compliance with the applicable provisions of this code.

R303.1.1 Building thermal envelope insulation. An *R*-value identification mark shall be applied by the manufacturer to each piece of *building thermal envelope* insulation 12 inches (305 mm) or greater in width. Alternately, the insulation installers shall provide a certification listing the type, manufacturer and *R*-value of insulation installed in each element of the *building thermal envelope*. For blown or sprayed insulation (fiberglass and cellulose), the initial installed thickness, settled thickness, settled *R*-value, installed density, coverage area and number of bags installed shall be *listed* on the certification. For sprayed polyurethane foam (SPF) insulation, the installed thickness of the areas covered and *R*-value of installed thickness shall be *listed* on the certification. The insulation installer shall sign, date and post the certification in a conspicuous location on the job site.

R303.1.1.1 Blown or sprayed roof/ceiling insulation. The thickness of blown-in or sprayed roof/ceiling insulation (fiberglass or cellulose) shall be written in inches (mm) on markers that are installed at least one for every 300 square feet (28 m²) throughout the attic space. The markers shall be affixed to the trusses or joists and marked with the minimum initial installed thickness with numbers a minimum of 1 inch (25 mm) in height. Each marker shall face the attic access opening. Spray polyurethane foam thickness and installed *R*-value shall be *listed* on certification provided by the insulation installer.

R303.1.2 Insulation mark installation. Insulating materials shall be installed such that the manufacturer's *R*-value mark is readily observable upon inspection.

R303.1.3 Fenestration product rating. *U*-factors of fenestration products (windows, doors and skylights) shall be determined in accordance with NFRC 100 by an accredited, independent laboratory, and labeled and certified by the manufacturer. Products lacking such a labeled *U*-factor shall be assigned a default *U*-factor from Table R303.1.3(1) or R303.1.3(2). The solar heat gain coefficient (SHGC) and *visible transmittance* (VT) of glazed fenestration products (windows, glazed doors and skylights) shall be determined in accordance with NFRC 200 by an accredited, independent laboratory, and labeled and certified by the manufacturer. Products lacking such a labeled SHGC or VT shall be assigned a default SHGC or VT from Table R303.1.3(3).

TABLE R303.1.3(1)
DEFAULT GLAZED FENESTRATION *U*-FACTOR

FRAME TYPE	SINGLE PANE	DOUBLE PANE	SKYLIGHT	
			Single	Double
Metal	1.20	0.80	2.00	1.30
Metal with Thermal Break	1.10	0.65	1.90	1.10
Nonmetal or Metal Clad	0.95	0.55	1.75	1.05
Glazed Block	0.60			

TABLE R303.1.3(2)
DEFAULT DOOR *U*-FACTORS

DOOR TYPE	*U*-FACTOR
Uninsulated Metal	1.20
Insulated Metal	0.60
Wood	0.50
Insulated, nonmetal edge, max 45% glazing, any glazing double pane	0.35

TABLE R303.1.3(3)
DEFAULT GLAZED FENESTRATION SHGC AND VT

	SINGLE GLAZED		DOUBLE GLAZED		GLAZED BLOCK
	Clear	Tinted	Clear	Tinted	
SHGC	0.8	0.7	0.7	0.6	0.6
VT	0.6	0.3	0.6	0.3	0.6

R303.1.4 Insulation product rating. The thermal resistance (*R*-value) of insulation shall be determined in accordance with the U.S. Federal Trade Commission *R*-value rule (CFR Title 16, Part 460) in units of h × ft^2 × °F/Btu at a mean temperature of 75°F (24°C).

R303.2 Installation. All materials, systems and equipment shall be installed in accordance with the manufacturer's installation instructions and the *International Building Code* or *International Residential Code*, as applicable.

R303.2.1 Protection of exposed foundation insulation. Insulation applied to the exterior of basement walls, crawlspace walls and the perimeter of slab-on-grade floors shall have a rigid, opaque and weather-resistant protective covering to prevent the degradation of the insulation's thermal performance. The protective covering shall cover the exposed exterior insulation and extend a minimum of 6 inches (153 mm) below grade.

R303.3 Maintenance information. Maintenance instructions shall be furnished for equipment and systems that require preventive maintenance. Required regular maintenance actions shall be clearly stated and incorporated on a readily accessible label. The label shall include the title or publication number for the operation and maintenance manual for that particular model and type of product.

RESIDENTIAL ENERGY EFFICIENCY

SECTION R401
GENERAL

R401.1 Scope. This chapter applies to residential buildings.

R401.2 Compliance. Projects shall comply with Sections identified as "mandatory" and with either sections identified as "prescriptive" or the performance approach in Section R405.

R401.3 Certificate (Mandatory). A permanent certificate shall be completed and posted on or in the electrical distribution panel by the builder or registered design professional. The certificate shall not cover or obstruct the visibility of the circuit directory label, service disconnect label or other required labels. The certificate shall list the predominant *R*-values of insulation installed in or on ceiling/roof, walls, foundation (slab, *basement wall,* crawlspace wall and/or floor) and ducts outside conditioned spaces; *U*-factors for fenestration and the solar heat gain coefficient (SHGC) of fenestration, and the results from any required duct system and building envelope air leakage testing done on the building. Where there is more than one value for each component, the certificate shall list the value covering the largest area. The certificate shall list the types and efficiencies of heating, cooling and service water heating equipment. Where a gas-fired unvented room heater, electric furnace, or baseboard electric heater is installed in the residence, the certificate shall list "gas-fired unvented room heater," "electric furnace" or "baseboard electric heater," as appropriate. An efficiency shall not be *listed* for gas-fired unvented room heaters, electric furnaces or electric baseboard heaters.

SECTION R402
BUILDING THERMAL ENVELOPE

R402.1 General (Prescriptive). The *building thermal envelope* shall meet the requirements of Sections R402.1.1 through R402.1.4.

R402.1.1 Insulation and fenestration criteria. The *building thermal envelope* shall meet the requirements of Table R402.1.1 based on the climate zone specified in Chapter 3.

R402.1.2 *R*-value computation. Insulation material used in layers, such as framing cavity insulation and insulating sheathing, shall be summed to compute the component *R*-value. The manufacturer's settled *R*-value shall be used for

TABLE R402.1.1
INSULATION AND FENESTRATION REQUIREMENTS BY COMPONENT[a]

CLIMATE ZONE	FENESTRATION *U*-FACTOR[b]	SKYLIGHT[b] *U*-FACTOR	GLAZED FENESTRATION SHGC[b, e]	CEILING *R*-VALUE	WOOD FRAME WALL *R*-VALUE	MASS WALL *R*-VALUE[i]	FLOOR *R*-VALUE	BASEMENT[c] WALL *R*-VALUE	SLAB[d] *R*-VALUE & DEPTH	CRAWL SPACE[c] WALL *R*-VALUE
1	NR	0.75	0.25	30	13	3/4	13	0	0	0
2	0.40	0.65	0.25	38	13	4/6	13	0	0	0
3	0.35	0.55	0.25	38	20 or 13+5[h]	8/13	19	5/13[f]	0	5/13
4 except Marine	0.35	0.55	0.40	49	20 or 13+5[h]	8/13	19	10 /13	10, 2 ft	10/13
5 and Marine 4	0.32	0.55	NR	49	20 or 13+5[h]	13/17	30[g]	15/19	10, 2 ft	15/19
6	0.32	0.55	NR	49	20+5 or 13+10[h]	15/20	30[g]	15/19	10, 4 ft	15/19
7 and 8	0.32	0.55	NR	49	20+5 or 13+10[h]	19/21	38[g]	15/19	10, 4 ft	15/19

For SI: 1 foot = 304.8 mm.

a. *R*-values are minimums. *U*-factors and SHGC are maximums. When insulation is installed in a cavity which is less than the label or design thickness of the insulation, the installed *R*-value of the insulation shall not be less than the *R*-value specified in the table.

b. The fenestration *U*-factor column excludes skylights. The SHGC column applies to all glazed fenestration. Exception: Skylights may be excluded from glazed fenestration SHGC requirements in Climate Zones 1 through 3 where the SHGC for such skylights does not exceed 0.30.

c. "15/19" means R-15 continuous insulation on the interior or exterior of the home or R-19 cavity insulation at the interior of the basement wall. "15/19" shall be permitted to be met with R-13 cavity insulation on the interior of the basement wall plus R-5 continuous insulation on the interior or exterior of the home. "10/13" means R-10 continuous insulation on the interior or exterior of the home or R-13 cavity insulation at the interior of the basement wall.

d. R-5 shall be added to the required slab edge *R*-values for heated slabs. Insulation depth shall be the depth of the footing or 2 feet, whichever is less in Climate Zones 1 through 3 for heated slabs.

e. There are no SHGC requirements in the Marine Zone.

f. Basement wall insulation is not required in warm-humid locations as defined by Figure R301.1 and Table R301.1.

g. Or insulation sufficient to fill the framing cavity, R-19 minimum.

h. First value is cavity insulation, second is continuous insulation or insulated siding, so "13+5" means R-13 cavity insulation plus R-5 continuous insulation or insulated siding. If structural sheathing covers 40 percent or less of the exterior, continuous insulation *R*-value shall be permitted to be reduced by no more than R-3 in the locations where structural sheathing is used – to maintain a consistent total sheathing thickness.

i. The second *R*-value applies when more than half the insulation is on the interior of the mass wall.

TABLE R402.1.3
EQUIVALENT *U*-FACTORS[a]

CLIMATE ZONE	FENESTRATION *U*-FACTOR	SKYLIGHT *U*-FACTOR	CEILING *U*-FACTOR	FRAME WALL *U*-FACTOR	MASS WALL *U*-FACTOR[b]	FLOOR *U*-FACTOR	BASEMENT WALL *U*-FACTOR	CRAWL SPACE WALL *U*-FACTOR
1	0.50	0.75	0.035	0.082	0.197	0.064	0.360	0.477
2	0.40	0.65	0.030	0.082	0.165	0.064	0.360	0.477
3	0.35	0.55	0.030	0.057	0.098	0.047	0.091[c]	0.136
4 except Marine	0.35	0.55	0.026	0.057	0.098	0.047	0.059	0.065
5 and Marine 4	0.32	0.55	0.026	0.057	0.082	0.033	0.050	0.055
6	0.32	0.55	0.026	0.048	0.060	0.033	0.050	0.055
7 and 8	0.32	0.55	0.026	0.048	0.057	0.028	0.050	0.055

a. Nonfenestration *U*-factors shall be obtained from measurement, calculation or an approved source.

b. When more than half the insulation is on the interior, the mass wall *U*-factors shall be a maximum of 0.17 in Climate Zone 1, 0.14 in Climate Zone 2, 0.12 in Climate Zone 3, 0.087 in Climate Zone 4 except Marine, 0.065 in Climate Zone 5 and Marine 4, and 0.057 in Climate Zones 6 through 8.

c. Basement wall *U*-factor of 0.360 in warm-humid locations as defined by Figure R301.1 and Table R301.1.

blown insulation. Computed *R*-values shall not include an *R*-value for other building materials or air films.

R402.1.3 *U*-factor alternative. An assembly with a *U*-factor equal to or less than that specified in Table R402.1.3 shall be permitted as an alternative to the *R*-value in Table R402.1.1.

R402.1.4 Total UA alternative. If the total *building thermal envelope* UA (sum of *U*-factor times assembly area) is less than or equal to the total UA resulting from using the *U*-factors in Table R402.1.3 (multiplied by the same assembly area as in the proposed building), the building shall be considered in compliance with Table R402.1.1. The UA calculation shall be done using a method consistent with the ASHRAE *Handbook of Fundamentals* and shall include the thermal bridging effects of framing materials. The SHGC requirements shall be met in addition to UA compliance.

R402.2 Specific insulation requirements (Prescriptive). In addition to the requirements of Section R402.1, insulation shall meet the specific requirements of Sections R402.2.1 through R402.2.12.

R402.2.1 Ceilings with attic spaces. When Section R402.1.1 would require R-38 in the ceiling, R-30 shall be deemed to satisfy the requirement for R-38 wherever the full height of uncompressed R-30 insulation extends over the wall top plate at the eaves. Similarly, R-38 shall be deemed to satisfy the requirement for R-49 wherever the full height of uncompressed R-38 insulation extends over the wall top plate at the eaves. This reduction shall not apply to the *U*-factor alternative approach in Section R402.1.3 and the total UA alternative in Section R402.1.4.

R402.2.2 Ceilings without attic spaces. Where Section R402.1.1 would require insulation levels above R-30 and the design of the roof/ceiling assembly does not allow sufficient space for the required insulation, the minimum required insulation for such roof/ceiling assemblies shall be R-30. This reduction of insulation from the requirements of Section R402.1.1 shall be limited to 500 square feet (46 m²) or 20 percent of the total insulated ceiling area, whichever is less. This reduction shall not apply to

the *U*-factor alternative approach in Section R402.1.3 and the total UA alternative in Section R402.1.4.

R402.2.3 Eave baffle. For air permeable insulations in vented attics, a baffle shall be installed adjacent to soffit and eave vents. Baffles shall maintain an opening equal or greater than the size of the vent. The baffle shall extend over the top of the attic insulation. The baffle shall be permitted to be any solid material.

R402.2.4 Access hatches and doors. Access doors from conditioned spaces to unconditioned spaces (e.g., attics and crawl spaces) shall be weatherstripped and insulated to a level equivalent to the insulation on the surrounding surfaces. Access shall be provided to all equipment that prevents damaging or compressing the insulation. A wood framed or equivalent baffle or retainer is required to be provided when loose fill insulation is installed, the purpose of which is to prevent the loose fill insulation from spilling into the living space when the attic access is opened, and to provide a permanent means of maintaining the installed *R*-value of the loose fill insulation.

R402.2.5 Mass walls. Mass walls for the purposes of this chapter shall be considered above-grade walls of concrete block, concrete, insulated concrete form (ICF), masonry cavity, brick (other than brick veneer), earth (adobe, compressed earth block, rammed earth) and solid timber/logs.

R402.2.6 Steel-frame ceilings, walls, and floors. Steel-frame ceilings, walls, and floors shall meet the insulation requirements of Table R402.2.6 or shall meet the *U*-factor requirements of Table R402.1.3. The calculation of the *U*-factor for a steel-frame envelope assembly shall use a series-parallel path calculation method.

R402.2.7 Floors. Floor insulation shall be installed to maintain permanent contact with the underside of the subfloor decking.

R402.2.8 Basement walls. Walls associated with conditioned basements shall be insulated from the top of the *basement wall* down to 10 feet (3048 mm) below grade or to the basement floor, whichever is less. Walls associated with unconditioned basements shall meet this requirement unless the floor overhead is insulated in accordance with Sections R402.1.1 and R402.2.7.

TABLE R402.2.6
STEEL-FRAME CEILING, WALL AND FLOOR INSULATION
(R-VALUE)

WOOD FRAME R-VALUE REQUIREMENT	COLD-FORMED STEEL EQUIVALENT R-VALUE[a]
Steel Truss Ceilings[b]	
R-30	R-38 or R-30 + 3 or R-26 + 5
R-38	R-49 or R-38 + 3
R-49	R-38 + 5
Steel Joist Ceilings[b]	
R-30	R-38 in 2 × 4 or 2 × 6 or 2 × 8 R-49 in any framing
R-38	R-49 in 2 × 4 or 2 × 6 or 2 × 8 or 2 × 10
Steel-Framed Wall 16″ O.C.	
R-13	R-13 + 4.2 or R-19 + 2.1 or R-21 + 2.8 or R-0 + 9.3 or R-15 + 3.8 or R-21 + 3.1
R-13 + 3	R-0 + 11.2 or R-13 + 6.1 or R-15 + 5.7 or R-19 + 5.0 or R-21 + 4.7
R-20	R-0 + 14.0 or R-13 + 8.9 or R-15 + 8.5 or R-19 + 7.8 or R-19 + 6.2 or R-21 + 7.5
R-20 + 5	R-13 + 12.7 or R-15 + 12.3 or R-19 + 11.6 or R-21 + 11.3 or R-25 + 10.9
R-21	R-0 + 14.6 or R-13 + 9.5 or R-15 + 9.1 or R-19 + 8.4 or R-21 + 8.1 or R-25 + 7.7
Steel Framed Wall, 24″ O.C	
R-13	R-0 + 9.3 or R-13 + 3.0 or R-15 + 2.4
R-13 + 3	R-0 + 11.2 or R-13 + 4.9 or R-15 + 4.3 or R-19 + 3.5 or R-21 + 3.1
R-20	R-0 + 14.0 or R-13 + 7.7 or R-15 + 7.1 or R-19 + 6.3 or R-21 + 5.9
R-20 + 5	R-13 + 11.5 or R-15 + 10.9 or R-19 + 10.1 or R-21 + 9.7 or R-25 + 9.1
R-21	R-0 + 14.6 or R-13 + 8.3 or R-15 + 7.7 or R-19 + 6.9 or R-21 + 6.5 or R-25 + 5.9
Steel Joist Floor	
R-13	R-19 in 2 × 6, or R-19 + 6 in 2 × 8 or 2 × 10
R-19	R-19 + 6 in 2 × 6, or R-19 + 12 in 2 × 8 or 2 × 10

a. Cavity insulation R-value is listed first, followed by continuous insulation R-value.
b. Insulation exceeding the height of the framing shall cover the framing.

R402.2.9 Slab-on-grade floors. Slab-on-grade floors with a floor surface less than 12 inches (305 mm) below grade shall be insulated in accordance with Table R402.1.1. The insulation shall extend downward from the top of the slab on the outside or inside of the foundation wall. Insulation located below grade shall be extended the distance provided in Table R402.1.1 by any combination of vertical insulation, insulation extending under the slab or insulation extending out from the building. Insulation extending away from the building shall be protected by pavement or by a minimum of 10 inches (254 mm) of soil. The top edge of the insulation installed between the *exterior wall* and the edge of the interior slab shall be permitted to be cut at a 45-degree (0.79 rad) angle away from the *exterior wall*. Slab-edge insulation is not required in jurisdictions designated by the *code official* as having a very heavy termite infestation.

R402.2.10 Crawl space walls. As an alternative to insulating floors over crawl spaces, crawl space walls shall be permitted to be insulated when the crawl space is not vented to the outside. Crawl space wall insulation shall be permanently fastened to the wall and extend downward from the floor to the finished grade level and then vertically and/or horizontally for at least an additional 24 inches (610 mm). Exposed earth in unvented crawl space foundations shall be covered with a continuous Class I vapor retarder in accordance with the *International Building Code* or *International Residential Code*, as applicable. All joints of the vapor retarder shall overlap by 6 inches (153 mm) and be sealed or taped. The edges of the vapor retarder shall extend at least 6 inches (153 mm) up the stem wall and shall be attached to the stem wall.

R402.2.11 Masonry veneer. Insulation shall not be required on the horizontal portion of the foundation that supports a masonry veneer.

R402.2.12 Sunroom insulation. All *sunrooms* enclosing conditioned space shall meet the insulation requirements of this code.

> **Exception:** For *sunrooms* with *thermal isolation*, and enclosing conditioned space, the following exceptions to the insulation *requirements* of this code shall apply:
>
> 1. The minimum ceiling insulation R-values shall be R-19 in Climate Zones 1 through 4 and R-24 in Climate Zones 5 through 8; and
>
> 2. The minimum wall R-value shall be R-13 in all climate zones. Wall(s) separating a *sunroom* with a *thermal isolation* from *conditioned space* shall meet the *building thermal envelope* requirements of this code.

R402.3 Fenestration (Prescriptive). In addition to the requirements of Section R402, fenestration shall comply with Sections R402.3.1 through R402.3.6.

R402.3.1 U-factor. An area-weighted average of fenestration products shall be permitted to satisfy the U-factor requirements.

R402.3.2 Glazed fenestration SHGC. An area-weighted average of fenestration products more than 50-percent glazed shall be permitted to satisfy the SHGC requirements.

R402.3.3 Glazed fenestration exemption. Up to 15 square feet (1.4 m²) of glazed fenestration per dwelling unit shall be permitted to be exempt from U-factor and SHGC requirements in Section R402.1.1. This exemption shall not apply to the U-factor alternative approach in Section R402.1.3 and the Total UA alternative in Section R402.1.4.

R402.3.4 Opaque door exemption. One side-hinged opaque door assembly up to 24 square feet (2.22 m²) in area is exempted from the U-factor requirement in Section R402.1.1. This exemption shall not apply to the U-factor alternative approach in Section R402.1.3 and the total UA alternative in Section R402.1.4.

R402.3.5 Sunroom *U*-factor. All *sunrooms* enclosing conditioned space shall meet the fenestration requirements of this code.

Exception: For *sunrooms* with *thermal isolation* and enclosing conditioned space, in Climate Zones 4 through 8, the following exceptions to the fenestration requirements of this code shall apply:

1. The maximum fenestration *U*-factor shall be 0.45; and

2. The maximum skylight *U*-factor shall be 0.70. New fenestration separating the *sunroom* with *thermal isolation* from *conditioned space* shall meet the *building thermal envelope* requirements of this code.

R402.3.6 Replacement fenestration. Where some or all of an existing fenestration unit is replaced with a new fenestration product, including sash and glazing, the replacement fenestration unit shall meet the applicable requirements for *U*-factor and SHGC in Table R402.1.1.

TABLE R402.4.1.1
AIR BARRIER AND INSULATION INSTALLATION

COMPONENT	CRITERIA[a]
Air barrier and thermal barrier	A continuous air barrier shall be installed in the building envelope. Exterior thermal envelope contains a continuous air barrier. Breaks or joints in the air barrier shall be sealed. Air-permeable insulation shall not be used as a sealing material.
Ceiling/attic	The air barrier in any dropped ceiling/soffit shall be aligned with the insulation and any gaps in the air barrier sealed. Access openings, drop down stair or knee wall doors to unconditioned attic spaces shall be sealed.
Walls	Corners and headers shall be insulated and the junction of the foundation and sill plate shall be sealed. The junction of the top plate and top of exterior walls shall be sealed. Exterior thermal envelope insulation for framed walls shall be installed in substantial contact and continuous alignment with the air barrier. Knee walls shall be sealed.
Windows, skylights and doors	The space between window/door jambs and framing and skylights and framing shall be sealed.
Rim joists	Rim joists shall be insulated and include the air barrier.
Floors (including above-garage and cantilevered floors)	Insulation shall be installed to maintain permanent contact with underside of subfloor decking. The air barrier shall be installed at any exposed edge of insulation.
Crawl space walls	Where provided in lieu of floor insulation, insulation shall be permanently attached to the crawlspace walls. Exposed earth in unvented crawl spaces shall be covered with a Class I vapor retarder with overlapping joints taped.
Shafts, penetrations	Duct shafts, utility penetrations, and flue shafts opening to exterior or unconditioned space shall be sealed.
Narrow cavities	Batts in narrow cavities shall be cut to fit, or narrow cavities shall be filled by insulation that on installation readily conforms to the available cavity space.
Garage separation	Air sealing shall be provided between the garage and conditioned spaces.
Recessed lighting	Recessed light fixtures installed in the building thermal envelope shall be air tight, IC rated, and sealed to the drywall.
Plumbing and wiring	Batt insulation shall be cut neatly to fit around wiring and plumbing in exterior walls, or insulation that on installation readily conforms to available space shall extend behind piping and wiring.
Shower/tub on exterior wall	Exterior walls adjacent to showers and tubs shall be insulated and the air barrier installed separating them from the showers and tubs.
Electrical/phone box on exterior walls	The air barrier shall be installed behind electrical or communication boxes or air sealed boxes shall be installed.
HVAC register boots	HVAC register boots that penetrate building thermal envelope shall be sealed to the sub-floor or drywall.
Fireplace	An air barrier shall be installed on fireplace walls. Fireplaces shall have gasketed doors.

a. In addition, inspection of log walls shall be in accordance with the provisions of ICC-400.

R402.4 Air leakage (Mandatory). The building thermal envelope shall be constructed to limit air leakage in accordance with the requirements of Sections R402.4.1 through R402.4.4.

R402.4.1 Building thermal envelope. The *building thermal envelope* shall comply with Sections R402.4.1.1 and R402.4.1.2. The sealing methods between dissimilar materials shall allow for differential expansion and contraction.

R402.4.1.1 Installation. The components of the *building thermal envelope* as listed in Table R402.4.1.1 shall be installed in accordance with the manufacturer's instructions and the criteria listed in Table R402.4.1.1, as applicable to the method of construction. Where required by the *code official*, an *approved* third party shall inspect all components and verify compliance.

R402.4.1.2 Testing. The building or dwelling unit shall be tested and verified as having an air leakage rate of not exceeding 5 air changes per hour in Climate Zones 1 and 2, and 3 air changes per hour in Climate Zones 3 through 8. Testing shall be conducted with a blower door at a pressure of 0.2 inches w.g. (50 Pascals). Where required by the *code official*, testing shall be conducted by an *approved* third party. A written report of the results of the test shall be signed by the party conducting the test and provided to the *code official*. Testing shall be performed at any time after creation of all penetrations of the *building thermal envelope*.

During testing:

1. Exterior windows and doors, fireplace and stove doors shall be closed, but not sealed, beyond the intended weatherstripping or other infiltration control measures;

2. Dampers including exhaust, intake, makeup air, backdraft and flue dampers shall be closed, but not sealed beyond intended infiltration control measures;

3. Interior doors, if installed at the time of the test, shall be open;

4. Exterior doors for continuous ventilation systems and heat recovery ventilators shall be closed and sealed;

5. Heating and cooling systems, if installed at the time of the test, shall be turned off; and

6. Supply and return registers, if installed at the time of the test, shall be fully open.

R402.4.2 Fireplaces. New wood-burning fireplaces shall have tight-fitting flue dampers and outdoor combustion air.

R402.4.3 Fenestration air leakage. Windows, skylights and sliding glass doors shall have an air infiltration rate of no more than 0.3 cfm per square foot (1.5 L/s/m^2), and swinging doors no more than 0.5 cfm per square foot (2.6 L/s/m^2), when tested according to NFRC 400 or AAMA/WDMA/CSA 101/I.S.2/A440 by an accredited, independent laboratory and *listed* and *labeled* by the manufacturer.

Exception: Site-built windows, skylights and doors.

R402.4.4 Recessed lighting. Recessed luminaires installed in the *building thermal envelope* shall be sealed to limit air leakage between conditioned and unconditioned spaces. All recessed luminaires shall be IC-rated and *labeled* as having an air leakage rate not more than 2.0 cfm (0.944 L/s) when tested in accordance with ASTM E 283 at a 1.57 psf (75 Pa) pressure differential. All recessed luminaires shall be sealed with a gasket or caulk between the housing and the interior wall or ceiling covering.

R402.5 Maximum fenestration *U*-factor and SHGC (Mandatory). The area-weighted average maximum fenestration *U*-factor permitted using tradeoffs from Section R402.1.4 or R405 shall be 0.48 in Climate Zones 4 and 5 and 0.40 in Climate Zones 6 through 8 for vertical fenestration, and 0.75 in Climate Zones 4 through 8 for skylights. The area-weighted average maximum fenestration SHGC permitted using tradeoffs from Section R405 in Climate Zones 1 through 3 shall be 0.50.

SECTION R403
SYSTEMS

R403.1 Controls (Mandatory). At least one thermostat shall be provided for each separate heating and cooling system.

R403.1.1 Programmable thermostat. Where the primary heating system is a forced-air furnace, at least one thermostat per dwelling unit shall be capable of controlling the heating and cooling system on a daily schedule to maintain different temperature set points at different times of the day. This thermostat shall include the capability to set back or temporarily operate the system to maintain *zone* temperatures down to 55°F (13°C) or up to 85°F (29°C). The thermostat shall initially be programmed with a heating temperature set point no higher than 70°F (21°C) and a cooling temperature set point no lower than 78°F (26°C).

R403.1.2 Heat pump supplementary heat (Mandatory). Heat pumps having supplementary electric-resistance heat shall have controls that, except during defrost, prevent supplemental heat operation when the heat pump compressor can meet the heating load.

R403.2 Ducts. Ducts and air handlers shall be in accordance with Sections R403.2.1 through R403.2.3.

R403.2.1 Insulation (Prescriptive). Supply ducts in attics shall be insulated to a minimum of R-8. All other ducts shall be insulated to a minimum of R-6.

Exception: Ducts or portions thereof located completely inside the *building thermal envelope*.

R403.2.2 Sealing (Mandatory). Ducts, air handlers, and filter boxes shall be sealed. Joints and seams shall comply with either the *International Mechanical Code* or *International Residential Code*, as applicable.

Exceptions:

1. Air-impermeable spray foam products shall be permitted to be applied without additional joint seals.

2. Where a duct connection is made that is partially inaccessible, three screws or rivets shall be equally spaced on the exposed portion of the joint so as to prevent a hinge effect.

3. Continuously welded and locking-type longitudinal joints and seams in ducts operating at static pressures less than 2 inches of water column (500 Pa) pressure classification shall not require additional closure systems.

Duct tightness shall be verified by either of the following:

1. Postconstruction test: Total leakage shall be less than or equal to 4 cfm (113.3 L/min) per 100 square feet (9.29 m²) of conditioned floor area when tested at a pressure differential of 0.1 inches w.g. (25 Pa) across the entire system, including the manufacturer's air handler enclosure. All register boots shall be taped or otherwise sealed during the test.

2. Rough-in test: Total leakage shall be less than or equal to 4 cfm (113.3 L/min) per 100 square feet (9.29 m²) of conditioned floor area when tested at a pressure differential of 0.1 inches w.g. (25 Pa) across the system, including the manufacturer's air handler enclosure. All registers shall be taped or otherwise sealed during the test. If the air handler is not installed at the time of the test, total leakage shall be less than or equal to 3 cfm (85 L/min) per 100 square feet (9.29 m²) of conditioned floor area.

   Exception: The total leakage test is not required for ducts and air handlers located entirely within the building thermal envelope.

   R403.2.2.1 Sealed air handler. Air handlers shall have a manufacturer's designation for an air leakage of no more than 2 percent of the design air flow rate when tested in accordance with ASHRAE 193.

R403.2.3 Building cavities (Mandatory). Building framing cavities shall not be used as ducts or plenums.

R403.3 Mechanical system piping insulation (Mandatory). Mechanical system piping capable of carrying fluids above 105°F (41°C) or below 55°F (13°C) shall be insulated to a minimum of R-3.

R403.3.1 Protection of piping insulation. Piping insulation exposed to weather shall be protected from damage, including that caused by sunlight, moisture, equipment maintenance, and wind, and shall provide shielding from solar radiation that can cause degradation of the material. Adhesive tape shall not be permitted.

R403.4 Service hot water systems. Energy conservation measures for service hot water systems shall be in accordance with Sections R403.4.1 and R403.4.2.

R403.4.1 Circulating hot water systems (Mandatory). Circulating hot water systems shall be provided with an automatic or readily *accessible* manual switch that can turn off the hot-water circulating pump when the system is not in use.

R403.4.2 Hot water pipe insulation (Prescriptive). Insulation for hot water pipe with a minimum thermal resistance (*R*-value) of R-3 shall be applied to the following:

1. Piping larger than ³/₄ inch nominal diameter.

2. Piping serving more than one dwelling unit.

3. Piping from the water heater to kitchen outlets.

4. Piping located outside the conditioned space.

5. Piping from the water heater to a distribution manifold.

6. Piping located under a floor slab.

7. Buried piping.

8. Supply and return piping in recirculation systems other than demand recirculation systems.

9. Piping with run lengths greater than the maximum run lengths for the nominal pipe diameter given in Table R403.4.2.

All remaining piping shall be insulated to at least R-3 or meet the run length requirements of Table R403.4.2.

TABLE R403.4.2
MAXIMUM RUN LENGTH (feet)[a]

Nominal Pipe Diameter of Largest Diameter Pipe in the Run (inch)	³/₈	¹/₂	³/₄	> ³/₄
Maximum Run Length	30	20	10	5

For SI: 1 inch = 25.4 mm, 1 foot 304.8 mm.

a. Total length of all piping from the distribution manifold or the recirculation loop to a point of use.

R403.5 Mechanical ventilation (Mandatory). The building shall be provided with ventilation that meets the requirements

TABLE R403.5.1
MECHANICAL VENTILATION SYSTEM FAN EFFICACY

FAN LOCATION	AIR FLOW RATE MINIMUM (CFM)	MINIMUM EFFICACY (CFM/WATT)	AIR FLOW RATE MAXIMUM (CFM)
Range hoods	Any	2.8 cfm/watt	Any
In-line fan	Any	2.8 cfm/watt	Any
Bathroom, utility room	10	1.4 cfm/watt	< 90
Bathroom, utility room	90	2.8 cfm/watt	Any

For SI: 1 cfm = 28.3 L/min.

of the *International Residential Code* or *International Mechanical Code*, as applicable, or with other approved means of ventilation. Outdoor air intakes and exhausts shall have automatic or gravity dampers that close when the ventilation system is not operating.

R403.5.1 Whole-house mechanical ventilation system fan efficacy. Mechanical ventilation system fans shall meet the efficacy requirements of Table R403.5.1.

Exception: Where mechanical ventilation fans are integral to tested and listed HVAC equipment, they shall be powered by an electronically commutated motor.

R403.6 Equipment Sizing (Mandatory). Heating and cooling equipment shall be sized in accordance with ACCA Manual S based on building loads calculated in accordance with ACCA Manual J or other *approved* heating and cooling calculation methodologies.

R403.7 Systems serving multiple dwelling units (Mandatory). Systems serving multiple dwelling units shall comply with Sections C403 and C404 of the IECC—Commercial Provisions in lieu of Section R403.

R403.8 Snow melt system controls (Mandatory). Snow- and ice-melting systems, supplied through energy service to the building, shall include automatic controls capable of shutting off the system when the pavement temperature is above 50°F, and no precipitation is falling and an automatic or manual control that will allow shutoff when the outdoor temperature is above 40°F.

R403.9 Pools and inground permanently installed spas (Mandatory). Pools and inground permanently installed spas shall comply with Sections R403.9.1 through R403.9.3.

R403.9.1 Heaters. All heaters shall be equipped with a readily *accessible* on-off switch that is mounted outside of the heater to allow shutting off the heater without adjusting the thermostat setting. Gas-fired heaters shall not be equipped with constant burning pilot lights.

R403.9.2 Time switches. Time switches or other control method that can automatically turn off and on heaters and pumps according to a preset schedule shall be installed on all heaters and pumps. Heaters, pumps and motors that have built in timers shall be deemed in compliance with this requirement.

Exceptions:

1. Where public health standards require 24-hour pump operation.

2. Where pumps are required to operate solar- and waste-heat-recovery pool heating systems.

R403.9.3 Covers. Heated pools and inground permanently installed spas shall be provided with a vapor-retardant cover.

Exception: Pools deriving over 70 percent of the energy for heating from site-recovered energy, such as a heat pump or solar energy source computed over an operating season.

SECTION R404
ELECTRICAL POWER AND LIGHTING SYSTEMS (MANDATORY)

R404.1 Lighting equipment (Mandatory). A minimum of 75 percent of the lamps in permanently installed lighting fixtures shall be high-efficacy lamps or a minimum of 75 percent of the permanently installed lighting fixtures shall contain only high efficacy lamps.

Exception: Low-voltage lighting shall not be required to utilize high-efficiency lamps.

R404.1.1 Lighting equipment (Mandatory). Fuel gas lighting systems shall not have continuously burning pilot lights.

SECTION R405
SIMULATED PERFORMANCE ALTERNATIVE (PERFORMANCE)

R405.1 Scope. This section establishes criteria for compliance using simulated energy performance analysis. Such analysis shall include heating, cooling, and service water heating energy only.

R405.2 Mandatory requirements. Compliance with this section requires that the mandatory provisions identified in Section R401.2 be met. All supply and return ducts not completely inside the *building thermal envelope* shall be insulated to a minimum of R-6.

R405.3 Performance-based compliance. Compliance based on simulated energy performance requires that a proposed residence (*proposed design*) be shown to have an annual energy cost that is less than or equal to the annual energy cost of the *standard reference design*. Energy prices shall be taken from a source *approved* by the *code official*, such as the Department of Energy, Energy Information Administration's *State Energy Price and Expenditure Report. Code officials* shall be permitted to require time-of-use pricing in energy cost calculations.

Exception: The energy use based on source energy expressed in Btu or Btu per square foot of *conditioned floor area* shall be permitted to be substituted for the energy cost. The source energy multiplier for electricity shall be 3.16. The source energy multiplier for fuels other than electricity shall be 1.1.

R405.4 Documentation. Documentation of the software used for the performance design and the parameters for the building shall be in accordance with Sections R405.4.1 through R405.4.3.

R405.4.1 Compliance software tools. Documentation verifying that the methods and accuracy of the compliance software tools conform to the provisions of this section shall be provided to the *code official*.

R405.4.2 Compliance report. Compliance software tools shall generate a report that documents that the *proposed* *design* complies with Section R405.3. The compliance documentation shall include the following information:

1. Address or other identification of the residence;

2. An inspection checklist documenting the building component characteristics of the *proposed design* as listed in Table R405.5.2(1). The inspection checklist shall show results for both the *standard reference design* and the *proposed design,* and shall document all inputs entered by the user necessary to reproduce the results;

3. Name of individual completing the compliance report; and

4. Name and version of the compliance software tool.

Exception: Multiple orientations. When an otherwise identical building model is offered in multiple orientations, compliance for any orientation shall be permitted by documenting that the building meets the performance requirements in each of the four cardinal (north, east, south and west) orientations.

R405.4.3 Additional documentation. The *code official* shall be permitted to require the following documents:

1. Documentation of the building component characteristics of the *standard reference design.*

2. A certification signed by the builder providing the building component characteristics of the *proposed design* as given in Table R405.5.2(1).

3. Documentation of the actual values used in the software calculations for the *proposed design.*

R405.5 Calculation procedure. Calculations of the performance design shall be in accordance with Sections R405.5.1 and R405.5.2.

R405.5.1 General. Except as specified by this section, the *standard reference design* and *proposed design* shall be configured and analyzed using identical methods and techniques.

R405.5.2 Residence specifications. The *standard reference design* and *proposed design* shall be configured and analyzed as specified by Table R405.5.2(1). Table R405.5.2(1) shall include by reference all notes contained in Table R402.1.1.

R405.6 Calculation software tools. Calculation software, where used, shall be in accordance with Sections R405.6.1 through R405.6.3.

R405.6.1 Minimum capabilities. Calculation procedures used to comply with this section shall be software tools capable of calculating the annual energy consumption of all building elements that differ between the *standard reference design* and the *proposed design* and shall include the following capabilities:

1. Computer generation of the *standard reference design* using only the input for the *proposed design.* The calculation procedure shall not allow the user to directly modify the building component characteristics of the *standard reference design.*

2. Calculation of whole-building (as a single *zone*) sizing for the heating and cooling equipment in the *standard reference design* residence in accordance with Section R403.6.

3. Calculations that account for the effects of indoor and outdoor temperatures and part-load ratios on the performance of heating, ventilating and air-conditioning equipment based on climate and equipment sizing.

4. Printed *code official* inspection checklist listing each of the *proposed design* component characteristics from Table R405.5.2(1) determined by the analysis to provide compliance, along with their respective performance ratings (e.g., *R*-value, *U*-factor, SHGC, HSPF, AFUE, SEER, EF, etc.).

R405.6.2 Specific approval. Performance analysis tools meeting the applicable sections of Section R405 shall be permitted to be *approved.* Tools are permitted to be *approved* based on meeting a specified threshold for a jurisdiction. The *code official* shall be permitted to approve tools for a specified application or limited scope.

R405.6.3 Input values. When calculations require input values not specified by Sections R402, R403, R404 and R405, those input values shall be taken from an *approved* source.

TABLE R405.5.2(1)
SPECIFICATIONS FOR THE STANDARD REFERENCE AND PROPOSED DESIGNS

BUILDING COMPONENT	STANDARD REFERENCE DESIGN	PROPOSED DESIGN
Above-grade walls	Type: mass wall if proposed wall is mass; otherwise wood frame. Gross area: same as proposed *U*-factor: from Table R402.1.3 Solar absorptance = 0.75 Remittance = 0.90	As proposed As proposed As proposed As proposed As proposed
Basement and crawl space walls	Type: same as proposed Gross area: same as proposed *U*-factor: from Table R402.1.3, with insulation layer on interior side of walls.	As proposed As proposed As proposed
Above-grade floors	Type: wood frame Gross area: same as proposed *U*-factor: from Table R402.1.3	As proposed As proposed As proposed
Ceilings	Type: wood frame Gross area: same as proposed *U*-factor: from Table R402.1.3	As proposed As proposed As proposed
Roofs	Type: composition shingle on wood sheathing Gross area: same as proposed Solar absorptance = 0.75 Emittance = 0.90	As proposed As proposed As proposed As proposed
Attics	Type: vented with aperture = 1 ft² per 300 ft² ceiling area	As proposed
Foundations	Type: same as proposed foundation wall area above and below grade and soil characteristics: same as proposed.	As proposed As proposed
Doors	Area: 40 ft² Orientation: North *U*-factor: same as fenestration from Table R402.1.3.	As proposed As proposed As proposed
Glazing[a]	Total area[b] = (a) The proposed glazing area; where proposed glazing area is less than 15% of the conditioned floor area. (b) 15% of the conditioned floor area; where the proposed glazing area is 15% or more of the conditioned floor area. Orientation: equally distributed to four cardinal compass orientations (N, E, S & W). *U*-factor: from Table R402.1.3 SHGC: From Table R402.1.1 except that for climates with no requirement (NR) SHGC = 0.40 shall be used. Interior shade fraction: 0.92-(0.21 × SHGC for the standard reference design) External shading: none	As proposed As proposed As proposed As proposed 0.92-(0.21 × SHGC as proposed) As proposed
Skylights	None	As proposed
Thermally isolated sunrooms	None	As proposed

(continued)

TABLE R405.5.2(1)—continued
SPECIFICATIONS FOR THE STANDARD REFERENCE AND PROPOSED DESIGNS

BUILDING COMPONENT	STANDARD REFERENCE DESIGN	PROPOSED DESIGN
Air exchange rate	Air leakage rate of 5 air changes per hour in Climate Zones 1 and 2, and 3 air changes per hour in Climate Zones 3 through 8 at a pressure of 0.2 inches w.g (50 Pa). The mechanical ventilation rate shall be in addition to the air leakage rate and the same as in the proposed design, but no greater than $0.01 \times CFA + 7.5 \times (N_{br} + 1)$ where: CFA = conditioned floor area N_{br} = number of bedrooms Energy recovery shall not be assumed for mechanical ventilation.	For residences that are not tested, the same air leakage rate as the standard reference design. For tested residences, the measured air exchange rate[c]. The mechanical ventilation rate[d] shall be in addition to the air leakage rate and shall be as proposed.
Mechanical ventilation	None, except where mechanical ventilation is specified by the proposed design, in which case: Annual vent fan energy use: $kWh/yr = 0.03942 \times CFA + 29.565 \times (N_{br} + 1)$ where: CFA = conditioned floor area N_{br} = number of bedrooms	As proposed
Internal gains	$IGain = 17,900 + 23.8 \times CFA + 4104 \times N_{br}$ (Btu/day per dwelling unit)	Same as standard reference design.
Internal mass	An internal mass for furniture and contents of 8 pounds per square foot of floor area.	Same as standard reference design, plus any additional mass specifically designed as a thermal storage element[e] but not integral to the building envelope or structure.
Structural mass	For masonry floor slabs, 80% of floor area covered by R-2 carpet and pad, and 20% of floor directly exposed to room air. For masonry basement walls, as proposed, but with insulation required by Table R402.1.3 located on the interior side of the walls For other walls, for ceilings, floors, and interior walls, wood frame construction	As proposed As proposed As proposed
Heating systems[f, g]	As proposed for other than electric heating without a heat pump. Where the proposed design utilizes electric heating without a heat pump the standard reference design shall be an air source heat pump meeting the requirements of Section R403 of the IECC—Commercial Provisions. Capacity: sized in accordance with Section R403.6	As proposed
Cooling systems[f, h]	As proposed Capacity: sized in accordance with Section R403.6.	As proposed
Service water Heating[f, g, h, i]	As proposed Use: same as proposed design	As proposed $gal/day = 30 + (10 \times N_{br})$
Thermal distribution systems		Thermal distribution system efficiency shall be as tested or as specified in Table R405.5.2(2) if not tested. Duct insulation shall be as proposed.
Thermostat	Type: Manual, cooling temperature setpoint = 75°F; Heating temperature setpoint = 72°F	Same as standard reference

(continued)

TABLE R405.5.2(1)—continued
SPECIFICATIONS FOR THE STANDARD REFERENCE AND PROPOSED DESIGNS

For SI: 1 square foot = 0.93 m², 1 British thermal unit = 1055 J, 1 pound per square foot = 4.88 kg/m², 1 gallon (U.S.) = 3.785 L,
°C = (°F-3)/1.8, 1 degree = 0.79 rad.

a. Glazing shall be defined as sunlight-transmitting fenestration, including the area of sash, curbing or other framing elements, that enclose conditioned space. Glazing includes the area of sunlight-transmitting fenestration assemblies in walls bounding conditioned basements. For doors where the sunlight-transmitting opening is less than 50 percent of the door area, the glazing area is the sunlight transmitting opening area. For all other doors, the glazing area is the rough frame opening area for the door including the door and the frame.

b. For residences with conditioned basements, R-2 and R-4 residences and townhouses, the following formula shall be used to determine glazing area:

$$AF = A_s \times FA \times F$$

where:

AF = Total glazing area.

A_s = Standard reference design total glazing area.

FA = (Above-grade thermal boundary gross wall area)/(above-grade boundary wall area + 0.5 × below-grade boundary wall area).

F = (Above-grade thermal boundary wall area)/(above-grade thermal boundary wall area + common wall area) or 0.56, whichever is greater.

and where:

Thermal boundary wall is any wall that separates conditioned space from unconditioned space or ambient conditions.

Above-grade thermal boundary wall is any thermal boundary wall component not in contact with soil.

Below-grade boundary wall is any thermal boundary wall in soil contact.

Common wall area is the area of walls shared with an adjoining dwelling unit.

L and CFA are in the same units.

c. Where required by the *code official,* testing shall be conducted by an *approved* party. Hourly calculations as specified in the ASHRAE *Handbook of Fundamentals,* or the equivalent shall be used to determine the energy loads resulting from infiltration.

d. The combined air exchange rate for infiltration and mechanical ventilation shall be determined in accordance with Equation 43 of 2001 ASHRAE *Handbook of Fundamentals,* page 26.24 and the "Whole-house Ventilation" provisions of 2001 ASHRAE *Handbook of Fundamentals*, page 26.19 for intermittent mechanical ventilation.

e. Thermal storage element shall mean a component not part of the floors, walls or ceilings that is part of a passive solar system, and that provides thermal storage such as enclosed water columns, rock beds, or phase-change containers. A thermal storage element must be in the same room as fenestration that faces within 15 degrees (0.26 rad) of true south, or must be connected to such a room with pipes or ducts that allow the element to be actively charged.

f. For a proposed design with multiple heating, cooling or water heating systems using different fuel types, the applicable standard reference design system capacities and fuel types shall be weighted in accordance with their respective loads as calculated by accepted engineering practice for each equipment and fuel type present.

g. For a proposed design without a proposed heating system, a heating system with the prevailing federal minimum efficiency shall be assumed for both the standard reference design and proposed design.

h. For a proposed design home without a proposed cooling system, an electric air conditioner with the prevailing federal minimum efficiency shall be assumed for both the standard reference design and the proposed design.

i. For a proposed design with a nonstorage-type water heater, a 40-gallon storage-type water heater with the prevailing federal minimum energy factor for the same fuel as the predominant heating fuel type shall be assumed. For the case of a proposed design without a proposed water heater, a 40-gallon storage-type water heater with the prevailing federal minimum efficiency for the same fuel as the predominant heating fuel type shall be assumed for both the proposed design and standard reference design.

TABLE R405.5.2(2)
DEFAULT DISTRIBUTION SYSTEM EFFICIENCIES FOR PROPOSED DESIGNS[a]

DISTRIBUTION SYSTEM CONFIGURATION AND CONDITION	FORCED AIR SYSTEMS	HYDRONIC SYSTEMS[b]
Distribution system components located in unconditioned space	—	0.95
Untested distribution systems entirely located in conditioned space[c]	0.88	1
"Ductless" systems[d]	1	—

For SI: 1 cubic foot per minute = 0.47 L/s, 1 square foot = 0.093m², 1 pound per square inch = 6895 Pa, 1 inch water gauge = 1250 Pa.

a. Default values given by this table are for untested distribution systems, which must still meet minimum requirements for duct system insulation.

b. Hydronic systems shall mean those systems that distribute heating and cooling energy directly to individual spaces using liquids pumped through closed-loop piping and that do not depend on ducted, forced airflow to maintain space temperatures.

c. Entire system in conditioned space shall mean that no component of the distribution system, including the air-handler unit, is located outside of the conditioned space.

d. Ductless systems shall be allowed to have forced airflow across a coil but shall not have any ducted airflow external to the manufacturer's air-handler enclosure.

CHAPTER 5 [RE]

REFERENCED STANDARDS

This chapter lists the standards that are referenced in various sections of this document. The standards are listed herein by the promulgating agency of the standard, the standard identification, the effective date and title, and the section or sections of this document that reference the standard. The application of the referenced standards shall be as specified in Section R106.

AAMA

American Architectural Manufacturers Association
1827 Walden Office Square
Suite 550
Schaumburg, IL 60173-4268

Standard reference number	Title	Referenced in code section number
AAMA/WDMA/CSA 101/I.S.2/A C440—11	North American Fenestration Standard/ Specifications for Windows, Doors and Unit Skylights	R402.4.3

ACCA

Air Conditioning Contractors of America
2800 Shirlington Road, Suite 300
Arlington, VA 22206

Standard reference number	Title	Referenced in code section number
Manual J—11	Residential Load Calculation Eighth Edition	R403.6
Manual S—10	Residential Equipment Selection	R403.6

ASHRAE

American Society of Heating, Refrigerating and Air-Conditioning Engineers, Inc.
1791 Tullie Circle, NE
Atlanta, GA 30329-2305

Standard reference number	Title	Referenced in code section number
ASHRAE—2009	ASHRAE Handbook of Fundamentals	R402.1.4, Table R405.5.2(1)
ASHRAE 193—2010	Method of Test for Determining the Airtightness of HVAC Equipment	R403.2.2.1

ASTM

ASTM International
100 Barr Harbor Drive
West Conshohocken, PA 19428-2859

Standard reference number	Title	Referenced in code section number
E 283—04	Test Method for Determining the Rate of Air Leakage Through Exterior Windows, Curtain Walls and Doors Under Specified Pressure Differences Across the Specimen	R402.4.4

CSA

Canadian Standards Association
5060 Spectrum Way
Mississauga, Ontario, Canada L4W 5N6

Standard reference number	Title	Referenced in code section number
AAMA/WDMA/CSA 101/I.S.2/A440—11	North American Fenestration Standard/Specification for Windows, Doors and Unit Skylights.	R402.4.3

ICC

International Code Council, Inc.
500 New Jersey Avenue, NW
6th Floor
Washington, DC 20001

Standard reference number	Title	Referenced in code section number
IBC—12	International Building Code®	R201.3, R303.2, R402.2.10
ICC 400—12	Standard on the Design and Construction of Log Structures	Table R402.4.1.1
IFC—12	International Fire Code®	R201.3
IFGC—12	International Fuel Gas Code®	R201.3
IMC—12	International Mechanical Code®	R201.3, R403.2.2, R403.5
IPC—12	International Plumbing Code®	R201.3
IRC—12	International Residential Code®	R201.3, R303.2, R402.2.10, R403.2.2, R403.5

NFRC

National Fenestration Rating Council, Inc.
6305 Ivy Lane, Suite 140
Greenbelt, MD 20770

Standard reference number	Title	Referenced in code section number
100—2010	Procedure for Determining Fenestration Products U-factors	R303.1.3
200—2010	Procedure for Determining Fenestration Product Solar Heat Gain Coefficients and Visible Transmittance at Normal Incidence.	R303.1.3
400—2010	Procedure for Determining Fenestration Product Air Leakage	R402.4.3

US-FTC

United States-Federal Trade Commission
600 Pennsylvania Avenue NW
Washington, DC 20580

Standard reference number	Title	Referenced in code section number
CFR Title 16 (May 31, 2005)	R-value Rule	R303.1.4

WDMA

Window and Door Manufacturers Association
1400 East Touhy Avenue, Suite 470
Des Plaines, IL 60018

Standard reference number	Title	Referenced in code section number
AAMA/WDMA/CSA 101/I.S.2/A440—11	North American Fenestration Standard/Specification for Windows, Doors and Unit Skylights.	R402.4.3

INDEX [RE]

EDITORIAL CHANGES – SECOND PRINTING

Page R-29, Table R402.1.1: column 6, row 4 now reads . . . 13

EDITORIAL CHANGES – THIRD PRINTING

Page R-29, TABLE R402.1.1: column 6, row 4 now reads . . . 20 or 13+5

Page R-37, TABLE R405.5.2(1): column 2, row 10, lines 11 and 12 now reads . . .

Glazing[a]	Total area[b] =
	    (a) The proposed glazing area; where proposed glazing area is less than 15% of the conditioned floor area.
	    (b) 15% of the conditioned floor area; where the proposed glazing area is 15% or more of the conditioned floor area.
		As proposed

	Orientation: equally distributed to four cardinal compass orientations (N, E, S & W).	As proposed
	U-factor: from Table R402.1.3	As proposed
	SHGC: From Table R402.1.1 except that for climates with no requirement (NR) SHGC = 0.40 shall be used.	As proposed
	Interior shade fraction: 0.92-(0.21 × SHGC for the standard reference design)	0.92-(0.21 × SHGC as proposed)
	External shading: none	As proposed

Page R-42, NFRC: standard reference number 100—2009 now reads . . . 100—2010

Page R-42, NFRC: standard reference number 200—2009 now reads . . . 200—2010

Page R-42, NFRC: standard reference number 400—2009 now reads . . . 400 2010

EDITORIAL CHANGES – FOURTH PRINTING

Page R-5, Section R106.1.1: line 1 now reads . . . **R106.1.1 Conflicts**. Where conflicts occur between provi-

Page R-29, Table R402.1.1: column 6, line 4 now reads . . . 20 or 13+5[h]

INTERNATIONAL CODE COUNCIL®

People Helping People Build a Safer World™

Imagine...
enjoying **membership benefits** that help you stay competitive in today's tight job market.

Imagine... increasing your code knowledge and sharpening your job skills with special member discounts on:

- World-class training and **educational programs** to help you keep your competitive edge while earning CEUs and LUs
- The latest **code-related news** to enhance your career potential
- **Member Discounts** on code books, CDs, specialized publications, and training materials to help you stay on top of the latest code changes and become more valuable to your organization.

PLUS – having exclusive access to **code opinions** from experts to answer your code-related questions

Imagine... extra benefits such as **Member-Only privileged access** to peers in online discussion groups as well as access to the industry's leading periodical, the *Building Safety Journal Online®*.

Imagine... receiving these valuable discounts, benefits, and more by simply becoming a member of the nation's leading developer of building safety codes, the International Code Council®.

Imagine that!

Free code book(s) – New members receive a free book or set of books (depending on membership category) from the latest edition of the International Codes.

JOIN TODAY!
888-ICC-SAFE, x33804
member@iccsafe.org
www.iccsafe.org

11-04304

ICC EVALUATION SERVICE

Most Widely Accepted and Trusted

The leader in evaluating building products for code compliance & sustainable attributes

- Close to a century of experience in building product evaluation

- Most trusted and most widely accepted evaluation body

- In-house staff of licensed engineers

- Leader in innovative product evaluations

- Direct links to reports within code documents for easy access

Three signature programs:

1. ICC-ES Evaluation Report Program

2. Plumbing, Mechanical and Fuel Gas (PMG) Listing Program

3. Sustainable Attributes Verification and Evaluation (SAVE) Program

ICC-ES is a subsidiary of the International Code Council (ICC), the developer of the International Codes. Together, ICC-ES and ICC incorporate proven techniques and practices into codes and related products and services that foster safe and sustainable design and construction.

For more information, please contact us at es@icc-es.org or call 1.800.423.6587 (x42237).

www.icc-es.org

10-03910

INTERNATIONAL CODE COUNCIL

People Helping People Build a Safer World™

Maintain your sustainable building expertise with Green / Energy Training

Today's focus on green building and energy efficiency means code officials and manufacturers need to stay educated. ICC Training and Education helps to ensure mastery of green building and energy efficiency codes.

ICC is the world's premier organization for building safety and sustainability. Code enforcement officials, architects, engineers and contractors all turn to ICC for information, certification, training and guidance on code compliance and sustainability.

ICC offers Seminars, Institutes, and Webinars designed to keep code professionals up-to-speed on emerging issues. Continuing Education Units (CEUs) are awarded and recognized in many states.

Ensure that you understand the latest in green building and energy efficiency!

For more information on ICC training, visit www.iccsafe.org/Education or call 1-888-422-7233, ext. 33818.

11-04271

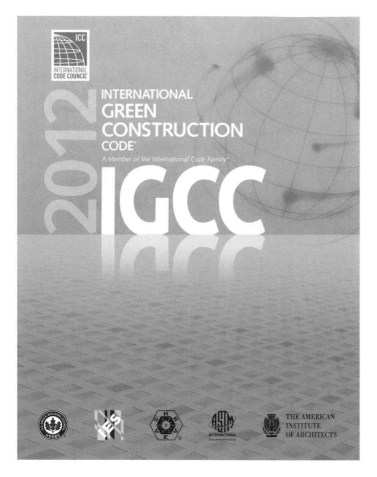

2012 INTERNATIONAL GREEN CONSTRUCTION CODE™ (IGCC™)

Release date: Spring 2012

In 2009, the International Code Council launched the development of a new model code focused on new and existing commercial buildings addressing affordable green building design and performance.

The IGCC provides the building industry with language that both broadens and strengthens building codes in a way that will accelerate the construction of high-performance, green buildings. It provides a vehicle for jurisdictions to regulate green for the design and performance of new and renovated buildings in a manner that is integrated with existing codes, allowing buildings to reap the rewards of improved design and construction practices.

Available Spring 2012

LEARN MORE! www.iccsafe.org/IGCC

COMMITTED TO BUILDING A GREEN AND SUSTAINABLE WORLD

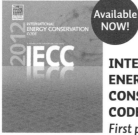

Available NOW!

INTERNATIONAL ENERGY CONSERVATION CODE®

First published: 1998

Encourages efficiency in envelope design, mechanical systems, lighting systems; and the use of new materials and techniques.

Available NOW!

ICC PERFORMANCE CODE®

First published: 2001

Presents regulations based on outcome rather than prescription to allow for new design methods and broader interpretation of the I-Codes.

Available NOW!

INTERNATIONAL EXISTING BUILDING CODE®

First published: 2003

Promotes the safe use of existing and historical structures to help restore neighborhoods and reduce waste.

Order your 2012 I-Codes today! 1-800-786-4452 | www.iccsafe.org/2012icodes

Expert Training Available! Visit www.iccsafe.org/training

INTERNATIONAL CODE COUNCIL®

People Helping People Build a Safer World™

11-04334